U0465250

中华思想文化术语传播工程

Key Concepts in
Chinese Thought and Culture

北京历史文化名城保护委员会办公室
北京市城市规划设计研究院　编

北京中轴线保护传承关键词

汉英对照

Beijing ¦¦ Central Axis ¦¦
Key Terms
(Chinese-English)

外语教学与研究出版社
FOREIGN LANGUAGE TEACHING AND RESEARCH PRESS
北京 BEIJING

图书在版编目（CIP）数据

北京中轴线保护传承关键词：汉英对照 ／ 北京历史文化名城保护委员会办公室，北京市城市规划设计研究院编． -- 北京：外语教学与研究出版社，2024.7

ISBN 978-7-5213-5248-1

Ⅰ.①北… Ⅱ.①北… ②北… Ⅲ.①城市规划－北京－普及读物－汉、英 Ⅳ.①TU984.21-49

中国国家版本馆 CIP 数据核字（2024）第 106907 号

地图审图号：京 S（2024）017

出 版 人	王 芳
责任编辑	王 琳
责任校对	赵璞玉
封面设计	李 高　彩奇风
版式设计	XXL Studio
出版发行	外语教学与研究出版社
社　　址	北京市西三环北路 19 号（100089）
网　　址	https://www.fltrp.com
印　　刷	天津善印科技有限公司
开　　本	710×1000　1/16
印　　张	10
字　　数	144 千字
版　　次	2024 年 7 月第 1 版
印　　次	2024 年 7 月第 1 次印刷
书　　号	ISBN 978-7-5213-5248-1
定　　价	59.00 元

如有图书采购需求，图书内容或印刷装订等问题，侵权、盗版书籍等线索，请拨打以下电话或关注官方服务号：
客服电话：400 898 7008
官方服务号：微信搜索并关注公众号"外研社官方服务号"
外研社购书网址：https://fltrp.tmall.com

物料号：352480001

教育部、国家语委重大文化工程
"中华思想文化术语传播工程"成果

BEIJING

编译团队

编写：北京历史文化名城保护委员会办公室
　　　北京市城市规划设计研究院

翻译：童孝华　邵亚楠　魏　侠

审订：柯鸿冈　黄希玲

EDITORIAL TEAM

Compilers　Office of Beijing Historic City Conservation Commission, Beijing Municipal Institute of City Planning & Design

Translator　TONG Xiaohua, SHAO Yanan, WEI Xia

Reviewers　Paul Ralph CROOK, HUANG Xiling

序言

习近平总书记指出，北京历史悠久，文脉绵长，是中华文明连续性、创新性、统一性、包容性、和平性的有力见证。

1982年1月，北京成为首批国家历史文化名城，同年3月编制完成的《北京城市建设总体规划方案》对保护中轴线提出明确要求。此后，中轴线一直是北京历史文化名城保护的重点内容。中轴线是元明清三朝都城规划和建设的核心，是北京老城空间布局和功能组织的统领，被誉为"北京老城的灵魂和脊梁"，是中国现存规模最为恢宏、保存最为完整的传统都城中轴线。

2023年1月，中国政府正式向联合国教科文组织申请将"北京中轴线——中国理想都城秩序的杰作"列入《世界遗产名录》。中轴线及其历史文化价值日益得到国内外越来越多的关注。

北京是全国政治中心、文化中心、国际交往中心和科技创新中心。为帮助国内外人士了解北京、了解中轴线，携手保护传承好这一人类共同的遗产，我们系统梳理与之相关的关键词，形成"历史文化价值""历史文化要素""世界遗产知识"等三方面共95个"关键词"，并根据有关国际组织文件、我国法律法规和法定规划、学术研究成果等撰写释义。

感谢"中华思想文化术语传播工程"支持出版《北京中轴线保护传承关键词（汉英对照）》。我们相信，本书及此前出版的《北京历史文化名城保护关键词（汉英对照）》将有助于更好地发挥北京作为历史古都和全国文化中心的优势，加强同全球各地的文化交流，共同推动文化繁荣发展、文化遗产保护、文明交流互鉴。

Preface

President Xi Jinping has remarked that Beijing, with its long history and profound cultural heritage, serves as a powerful testament to the continuity, innovation, unity, inclusiveness, and peaceful nature of the Chinese civilization.

In January 1982, Beijing was designated as one of the first batch of historic cities in China. The *Beijing Urban Development Master Plan*, finalized in March of the same year, specified requirements for the conservation of the city's Central Axis. Ever since, the Central Axis has been a focal point in the conservation efforts of Beijing's historic heritage. It is the benchmark of urban planning and construction during the Yuan, Ming and Qing dynasties, commanding the spatial organization and functional layout of Beijing's Old City. Celebrated as the "soul and backbone of the Old City", it is the grandest and most intact central axis of all that remain in traditional capitals in China.

In January 2023, the Chinese government formally submitted to UNESCO the dossier of "Beijing Central Axis: A Building Ensemble Exhibiting the Ideal Order of the Chinese Capital" for inscription on the World Heritage List. Since then, the Central Axis and its significance have gained growing recognition around the world and within China.

Beijing is the national political centre, cultural centre, centre for international exchanges, and centre for scientific discovery and technological innovation. To assist people both from home and abroad in better understanding Beijing and its Central Axis, we have assembled here 95 essential terms. These terms are categorized into three groups: Historic Value, Historic Elements, and Knowledge of World Heritage. The definitions are crafted based on international instruments, Chinese laws and regulations, statutory plans, and outcomes of academic research.

Our thanks go to the Key Concepts in Chinese Thought and Culture Translation and Communication Project for making this publication possible. We are confident that this book, along with the previously published bilingual version of *Conservation of the Historic City of Beijing: Key Terms*, will enhance Beijing's profile as both an ancient capital and a vibrant cultural centre, bolstering Beijing's efforts in international cultural exchange through global partnerships for conservation of cultural heritage, intercultural dialogue and mutual learning among civilizations.

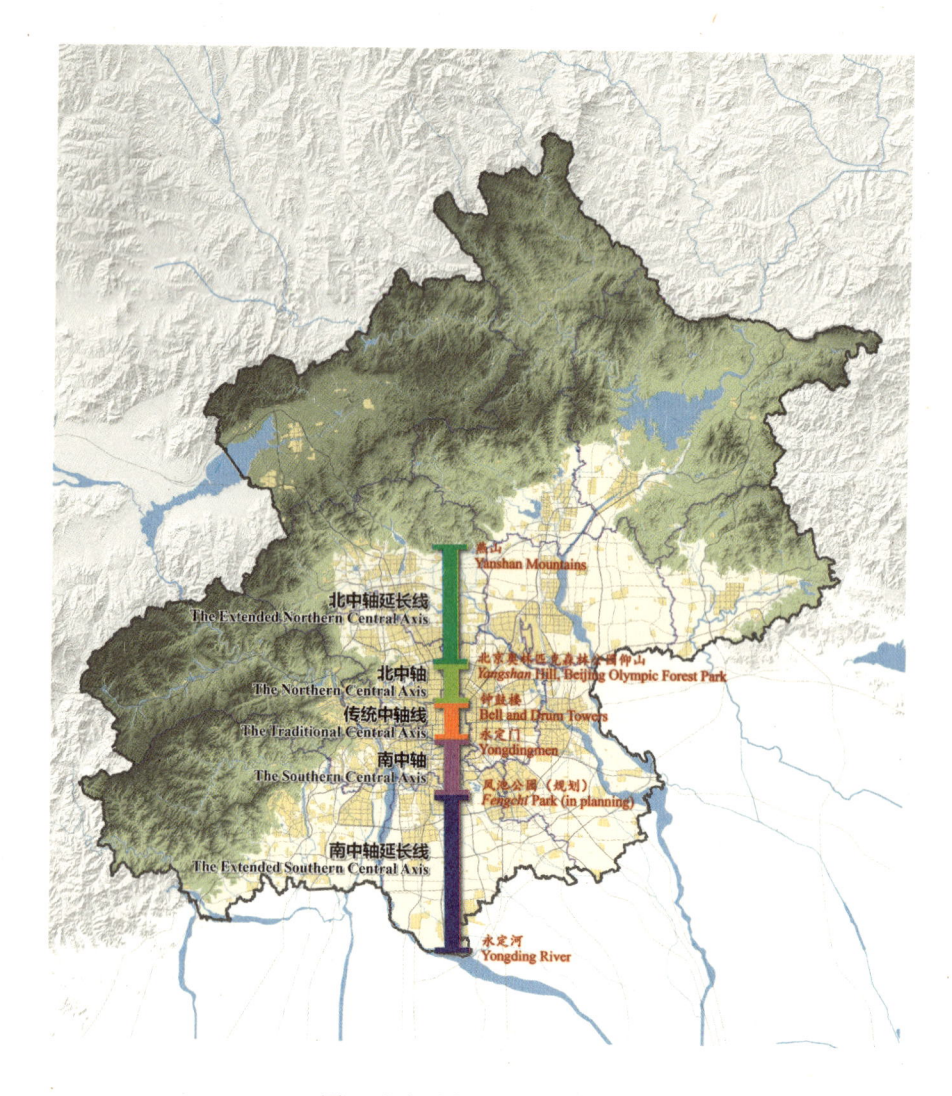

图 1. 北京城市中轴线示意图

Figure 1. The Central Axis of Beijing City

图 例

北京市行政边界
Boundary of Beijing City

北京市行政区界
Boundary of the Administrative Districts of Beijing

N

0 5 10   20   30公里

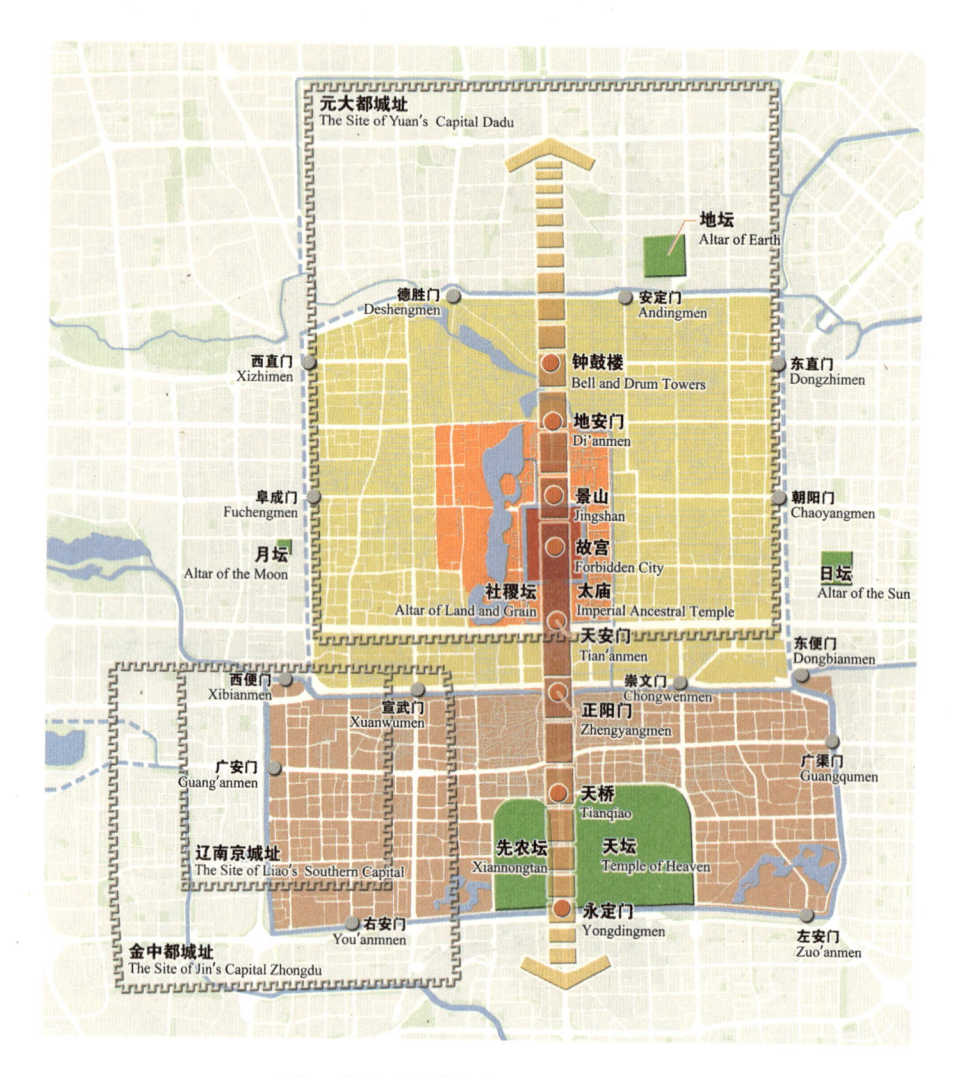

图2. 北京老城传统空间格局保护示意图
Figure 2. The Configuration of the Old City of Beijing

图例

- 宫城 The Palace City
- 皇城 The Imperial City
- 内城 The Inner City
- 外城 The Outer City
- 胡同 Hutongs
- 中轴线 The Central Axis
- 重要坛庙 Altars and Temples
- 绿地 Green Space
- 水系 Waters
- 暗沟 Hidden Water Channels
- 中轴线现存传统标志物 Nodes of the Central Axis
- 城门及角楼 City Gates and Corner Towers
- 辽金元都城址范围 Sites of the Capital City of Liao, Jin and Yuan Dynasties

0 0.25 0.5 1 1.5 公里

xi

Code	中文	English
ZA01	鼓楼—景山互眺	View between the Drum Tower and the Jinshan
ZA02	景山—故宫互眺	View between the Jingshan and the Forbidden City
ZA03	天安门—正阳门互眺	View between Tian'anmen and Zhengyangmen
ZA04	正阳门—永定门互眺	View between Zhengyangmen and Yongdingmen
ZA05	正阳门—祈年殿互眺	View between Zhengyangmen and the Hall of Prayer for Good Harvests
ZA06	永定门—祈年殿互眺	View between the Hall of Prayer for Good Harvests and Yongdingmen
ZA07	鼓楼望钟楼	The Bell Tower viewed from the Drum Tower
ZB01	银锭桥望钟楼	The Bell Tower viewed from the Yinding Bridge
ZB02	鼓楼东西大街望鼓楼	The Drum Tower viewed from Gulou East and West Streets
ZB03	地安门外大街望鼓楼	The Drum Tower viewed from Di'anmen Outer Street
ZB04	地安门路口望寿皇殿	The Hall of Imperial Longevity viewed from the crossroad of Di'anmen
ZB05	文津街望故宫西北角楼	The Northwest Corner Tower of the Forbidden City viewed from Wenjin Street
ZB06	东华门大街望东华门	Donghuamen viewed from Donghuamen Street
ZB07	前门东西大街望正阳门	Zhengyangmen viewed from Qianmen East and West Streets
ZB08	永定门东西大街望永定门	Yongdingmen viewed from Yongdingmen East and West Streets
ZB09	永定门公园望永定门	Yongdingmen viewed from Yongdingmen Park
ZB10	体育馆路望祈年殿	The Hall of Prayer for Good Harvests viewed from Tiyuguan Road
ZB11	祈年大街望祈年殿	The Hall of Prayer for Good Harvests viewed from Qinian Street
ZC01	钟楼望北中轴	The Northern Central Axis viewed from the Bell Tower
ZC02	永定门望南中轴	The Southern Central Axis viewed from Yongdingmen

图 3. 传统中轴线景观视廊示意图

Figure 3. Visual Corridors Concerning the Traditional Central Axis

图例
Legend

北京老城边界
Boundary of the Old City of Beijing

战略性景观标志物
Strategic Landmarks

景观视廊视线方向
View Direction of Visual Corridors

绿地
Green Space

水系
Waters

N

0 500 1000 2000米

目 录
Contents

第一篇　历史文化价值 001
Part I　Historic Value

Běijīng chéngshì zhōngzhóuxiàn
1. 北京城市中轴线 003
The Central Axis of Beijing City

chuántǒng zhōngzhóuxiàn
2. 传统中轴线 004
The Traditional Central Axis

tiānrén-héyī
3. 天人合一 005
Harmony Between Nature and Humanity

guānxiàng-shòushí
4. 观象授时 006
Observing Celestial Phenomena to Determine Time

xiàng tiān fǎ dì
5. 象天法地 ···················007
Emulating the Patterns of Heaven and Earth

biàn fāng zhèng wèi
6. 辨方正位 ···················008
Determining Directions and Establishing Proper Positions

yǐ zhōng wéi zūn
7. 以中为尊 ···················009
According Dignity to the Centre

jiànjí suíyóu
8. 建极绥猷 ···················010
Establishing the National Order and Implementing a Governance of Edification

yǔnzhí juézhōng
9. 允执厥中 ···················011
Holding Fast the Golden Mean

huáng jiàn yǒu jí
10. 皇建有极 ···················012
Extensively Establishing the National Order and Legal System

lǐ
11. 礼 ⋯⋯⋯⋯⋯⋯⋯⋯⋯⋯⋯⋯⋯⋯⋯⋯⋯⋯⋯⋯⋯⋯⋯⋯⋯⋯⋯⋯ 013
Li

yuè
12. 乐 ⋯⋯⋯⋯⋯⋯⋯⋯⋯⋯⋯⋯⋯⋯⋯⋯⋯⋯⋯⋯⋯⋯⋯⋯⋯⋯⋯⋯ 014
Yue

lǐyuè xiāngchéng
13. 礼乐相成 ⋯⋯⋯⋯⋯⋯⋯⋯⋯⋯⋯⋯⋯⋯⋯⋯⋯⋯⋯⋯⋯⋯⋯⋯ 015
Li and *Yue* Working in Synergy

zhōnghé zhī měi
14. 中和之美 ⋯⋯⋯⋯⋯⋯⋯⋯⋯⋯⋯⋯⋯⋯⋯⋯⋯⋯⋯⋯⋯⋯⋯⋯ 016
The Beauty of Harmony and Equilibrium

jìlǐ
15. 祭礼 ⋯⋯⋯⋯⋯⋯⋯⋯⋯⋯⋯⋯⋯⋯⋯⋯⋯⋯⋯⋯⋯⋯⋯⋯⋯⋯ 017
Sacrificial Ceremony

《Kǎogōng Jì》
16. 《考工记》 ⋯⋯⋯⋯⋯⋯⋯⋯⋯⋯⋯⋯⋯⋯⋯⋯⋯⋯⋯⋯⋯⋯⋯ 018
Kaogongji (*Book of Diverse Crafts*)

zuǒ zǔ yòu shè
17. 左祖右社 ⋯⋯⋯⋯⋯⋯⋯⋯⋯⋯⋯⋯⋯⋯⋯⋯⋯⋯⋯⋯⋯⋯⋯⋯ 019
Ancestral Temple on the Left and Altar of Land and Grain on the Right

miàn cháo hòushì
18. 面朝后市 ·· 020
Palace in the Front and Market in the Back

chuántǒng guīzhì
19. 传统规制 ·· 021
Traditional Specifications

móshù
20. 模数 ··· 022
Modularity

第二篇　历史文化要素　　　　　　　　　　　　　023
Part II　Historic Elements

Běijīng zhōngzhóuxiàn wénhuà yíchǎn
21. 北京中轴线文化遗产 ·· 025
Beijing Central Axis Cultural Heritage

Míng-Qīng Běijīngchéng
22. 明清北京城 ·· 026
The City of Beijing in the Ming and Qing Dynasties

nèichéng
23. 内城 ··· 027
The Inner City

wàichéng
24. 外城 028
The Outer City

gōngchéng
25. 宫城 029
The Palace City

huángchéng
26. 皇城 030
The Imperial City

sì chóng chéngkuò
27. 四重城廓 031
The Quadruple-Walled City Layout

qípán lùwǎng
28. 棋盘路网 032
The Chessboard Grid

jǐngguān shìláng
29. 景观视廊 033
Visual Corridors

jiēdào duìjǐng
30. 街道对景 034
Street View Corridors

xix

Zhōng Lóu
31. 钟楼 035
The Bell Tower

Gǔ Lóu
32. 鼓楼 036
The Drum Tower

Shíchà Hǎi
33. 什刹海 037
Shichahai

Wànníng Qiáo
34. 万宁桥 038
The Wanning Bridge

Dì'ān Mén
35. 地安门 039
Di'anmen (Gate of Earthly Peace)

Jǐng Shān
36. 景山 040
Jingshan

Shòuhuángdiàn jiànzhùqún
37. 寿皇殿建筑群 041
The Hall of Imperial Longevity Complex

Běishàng Mén
38. 北上门···042
Beishangmen (North Ascending Gate)

Gùgōng
39. 故宫···043
The Forbidden City

Duān Mén
40. 端门···044
Duanmen (Gate of Correct Deportment)

Tài Miào
41. 太庙···045
The Imperial Ancestral Temple

Shèjì Tán
42. 社稷坛··046
The Altar of Land and Grain

Tiān'ān Mén
43. 天安门··047
Tian'anmen (Gate of Heavenly Peace)

Tiān'ānmén huábiǎo
44. 天安门华表···048
Tian'anmen Huabiao Columns

Tiān'ānmén Guānlǐtái
45. 天安门观礼台 ·049
Tian'anmen Reviewing Stands

Nèijīnshuǐ Qiáo
46. 内金水桥 ·050
The Inner Golden Water Bridges

Wàijīnshuǐ Qiáo
47. 外金水桥 ·051
The Outer Golden Water Bridges

Qiānbù Láng
48. 千步廊 ·052
The Thousand-Step Galleries

Tiān'ānmén Guǎngchǎng
49. 天安门广场 ·053
Tian'anmen Square

Tiān'ānmén Guǎngchǎng guóqígān
50. 天安门广场国旗杆 ·054
The Flagpole on Tian'anmen Square

Rénmín Yīngxióng Jìniànbēi
51. 人民英雄纪念碑 ·055
The Monument to the People's Heroes

Máo Zhǔxí Jìniàntáng
52. 毛主席纪念堂···056
Chairman Mao Memorial Hall

Zhōngguó Guójiā Bówùguǎn
53. 中国国家博物馆···057
The National Museum of China

Rénmín Dàhuìtáng
54. 人民大会堂··058
The Great Hall of the People

Zhèngyáng Mén
55. 正阳门···059
Zhengyangmen

Zhèngyáng Qiáo
56. 正阳桥···060
The Zhengyang Bridge

Zhèngyáng Qiáo Páilou
57. 正阳桥牌楼··061
The Zhengyang Bridge Pailou

jūzhōng dàolù
58. 居中道路··062
Central Streets

xxiii

Qiánmén Dàjiē
59. 前门大街 ··· 063
Qianmen Street

dāngdāngchē
60. 铛铛车 ··· 064
Diang Diang Tram

zhōngzhóuxiàn nánduàn dàolù yícún
61. 中轴线南段道路遗存 ··· 065
Road Remains of the Southern Section of the Traditional Central Axis

Tiān Qiáo
62. 天桥 ··· 066
Tianqiao (Heavenly Bridge)

Zhèngyáng Qiáo Shūqú Jì Fāngbēi
63. 正阳桥疏渠记方碑 ··· 067
The Stele with Record of Dredging the Channels near the Zhengyang Bridge

Tiān Tán
64. 天坛 ··· 068
The Temple of Heaven

Xiānnóng Tán
65. 先农坛 ··· 069
Xiannongtan (Altar of the God of Agriculture)

Tiānshén Tán hé Dìqí Tán

66. 天神坛和地祇坛 ········· 070

The Altar of Celestial Gods and the Altar of Terrestrial Deities

yī mǔ sān fēn dì

67. 一亩三分地 ········· 071

One *Mu* and Three *Fens* of Land

Yǒngdìng Mén

68. 永定门 ········· 072

Yongdingmen

Yàndūn

69. 燕墩 ········· 073

Yandun

lǎozìhao

70. 老字号 ········· 074

Time-Honoured Brands

zhōng hé sháo yuè

71. 中和韶乐 ········· 075

Zhong He Shao Yue

bāyì wǔ

72. 八佾舞 ········· 076

Bayi Dance

第三篇　世界遗产知识　　　077
Part II　Knowledge of World Heritage

wénhuà yíchǎn
73. 文化遗产 ··· 079
Cultural Heritage

shìjiè yíchǎn
74. 世界遗产 ··· 080
World Heritage

shìjiè wénhuà yíchǎn
75. 世界文化遗产 ·· 081
World Cultural Heritage

shìjiè zìrán yíchǎn
76. 世界自然遗产 ·· 082
World Natural Heritage

shìjiè wénhuà hé zìrán hùnhé yíchǎn
77. 世界文化和自然混合遗产 ··· 083
World Mixed Cultural and Natural Heritage

《Shìjiè Yíchǎn Gōngyuē》
78.《世界遗产公约》··· 084
World Heritage Convention

《Shíshī 〈Shìjiè Yíchǎn Gōngyuē〉 Cāozuò Zhǐnán》
79.《实施〈世界遗产公约〉操作指南》⋯⋯⋯⋯⋯⋯⋯⋯⋯⋯085
Operational Guidelines for the Implementation of the World Heritage Convention

Shìjiè Yíchǎn Wěiyuánhuì
80. 世界遗产委员会⋯⋯⋯⋯⋯⋯⋯⋯⋯⋯⋯⋯⋯⋯⋯⋯⋯086
World Heritage Committee

Shìjiè Yíchǎn Zhōngxīn
81. 世界遗产中心⋯⋯⋯⋯⋯⋯⋯⋯⋯⋯⋯⋯⋯⋯⋯⋯⋯⋯087
World Heritage Centre

Guójì Gǔjì Yízhǐ Lǐshìhuì
82. 国际古迹遗址理事会⋯⋯⋯⋯⋯⋯⋯⋯⋯⋯⋯⋯⋯⋯⋯088
International Council on Monuments and Sites (ICOMOS)

《Shìjiè Yíchǎn Mínglù》
83.《世界遗产名录》⋯⋯⋯⋯⋯⋯⋯⋯⋯⋯⋯⋯⋯⋯⋯⋯⋯089
World Heritage List

tūchū pǔbiàn jiàzhí
84. 突出普遍价值⋯⋯⋯⋯⋯⋯⋯⋯⋯⋯⋯⋯⋯⋯⋯⋯⋯⋯090
Outstanding Universal Value (OUV)

shìjiè wénhuà yíchǎn jiàzhí biāozhǔn
85. 世界文化遗产价值标准⋯⋯⋯⋯⋯⋯⋯⋯⋯⋯⋯⋯⋯⋯091
Criteria for the Assessment of OUV of Cultural Heritage Sites

zhànlüè mùbiāo
86. 战略目标⋯⋯⋯⋯⋯⋯⋯⋯⋯⋯⋯⋯⋯⋯⋯⋯⋯⋯⋯⋯⋯⋯⋯⋯⋯⋯⋯⋯⋯093
Strategic Objectives

wánzhěngxìng
87. 完整性⋯⋯⋯⋯⋯⋯⋯⋯⋯⋯⋯⋯⋯⋯⋯⋯⋯⋯⋯⋯⋯⋯⋯⋯⋯⋯⋯⋯⋯⋯⋯094
Integrity

zhēnshíxìng
88. 真实性⋯⋯⋯⋯⋯⋯⋯⋯⋯⋯⋯⋯⋯⋯⋯⋯⋯⋯⋯⋯⋯⋯⋯⋯⋯⋯⋯⋯⋯⋯⋯095
Authenticity

yánxùxìng
89. 延续性⋯⋯⋯⋯⋯⋯⋯⋯⋯⋯⋯⋯⋯⋯⋯⋯⋯⋯⋯⋯⋯⋯⋯⋯⋯⋯⋯⋯⋯⋯⋯096
Continuity

zhěngtǐ bǎohù
90. 整体保护⋯⋯⋯⋯⋯⋯⋯⋯⋯⋯⋯⋯⋯⋯⋯⋯⋯⋯⋯⋯⋯⋯⋯⋯⋯⋯⋯⋯⋯⋯097
Comprehensive Conservation

yíchǎnqū
91. 遗产区⋯⋯⋯⋯⋯⋯⋯⋯⋯⋯⋯⋯⋯⋯⋯⋯⋯⋯⋯⋯⋯⋯⋯⋯⋯⋯⋯⋯⋯⋯⋯098
Boundaries for Effective Protection

huǎnchōngqū
92. 缓冲区⋯⋯⋯⋯⋯⋯⋯⋯⋯⋯⋯⋯⋯⋯⋯⋯⋯⋯⋯⋯⋯⋯⋯⋯⋯⋯⋯⋯⋯⋯⋯099
Buffer Zones

yíchǎn gòuchéng yàosù
93. 遗产构成要素···100
Elements

shìjiè yíchǎn de bàogào yǔ jiāncè
94. 世界遗产的报告与监测·······································101
Reporting and Monitoring of the World Heritage Sites

fǎnyìngxìng jiāncè
95. 反应性监测···102
Reactive Monitoring

附录 Appendix   103
    附录一　中国历史年代简表   103
    Appendix 1 A Brief Chronology of Chinese History
    附录二　中轴线演进年表   105
    Appendix 2 A Brief Chronology of the Evolution for the Traditional Central Axis

参考文献 Bibliography   112

索引 Index   115

第一篇
历史文化价值
Part I
Historic Value

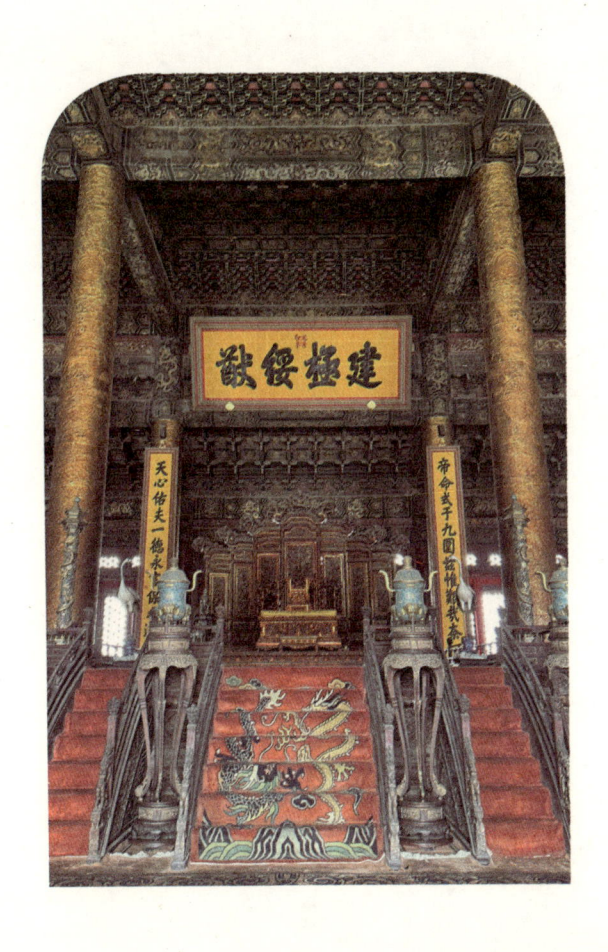

Běijīng chéngshì zhōngzhóuxiàn
1. 北京城市中轴线
The Central Axis of Beijing City

北京城市中轴线以传统中轴线为中心向南北延伸形成，北抵燕山山脉，南至北京大兴国际机场、永定河水系，是北京城市空间结构与功能组织的核心，可分为传统中轴线、北中轴、北中轴延长线、南中轴、南中轴延长线五段。

The Central Axis of Beijing City, with the traditional axis at its core, extends north to reach the Yanshan Mountains and stretches south to the Beijing Daxing International Airport and the Yongding River. This axis is the commanding element of Beijing City's urban spatial structure and functional organization. It can be divided into five sections: the Traditional Central Axis, the Northern Central Axis, the Extended Northern Central Axis, the Southern Central Axis, and the Extended Southern Central Axis.

chuántǒng zhōngzhóuxiàn

2. 传统中轴线

The Traditional Central Axis

　　元明清三朝都城规划和建设的核心，以及北京老城空间布局和功能组织的统领。传统中轴线始建于元代，发展完善于明清，居于北京老城中心区域、纵贯南北，北端自钟楼、鼓楼，向南经过万宁桥、景山、紫禁城、正阳门，南端至永定门，全长约7.8千米，它是中国现存规模最为恢宏、保存最为完整的传统都城中轴线，是都城中轴线的典范之作。

The Traditional Central Axis refers to the Central Axis at the heart of the capital's planning and construction during the Yuan, Ming and Qing dynasties. It is the commanding element in the spatial organization and functional layout of Beijing's Old City. Initially constructed in the Yuan Dynasty, the traditional axis was refined and reached its zenith during the Ming and Qing dynasties. It is situated in the middle of Beijing's Old City, extending in a straight line from north to south. The axis begins at the northern end with the Bell and Drum Towers, stretches southward through landmarks like the Wanning Bridge, Jingshan, the Forbidden City, and Zhengyangmen, and concludes at Yongdingmen, covering a length of 7.8 kilometres. It is the grandest and most intact central axis of all that remain in traditional capitals in China, and it is considered an exemplar in this regard.

tiānrén-héyī

3. 天人合一

Harmony Between Nature and Humanity

  一种认为天地人相通的世界观和思维方式。它旨在强调天地和人之间的整体性和内在联系，突出了天对于人或人事的根源性意义，表现了人在与天的联系中寻求生命、秩序与价值基础的努力。天人合一的思想体现在传统中轴线的选址和布局等诸多方面。

This term conveys a worldview and a way of thinking which hold that heaven, earth and man are interconnected. It underscores the symbiotic and intrinsic connection between them, and accentuates the pivotal role that heaven plays in human endeavours, and delineates the efforts of humanity to seek existence, organization, and ethics in concert with nature. This perspective is illustrated in the choice of location and design of the Traditional Central Axis.

guānxiàng-shòushí

4. 观象授时

Observing Celestial Phenomena to Determine Time

通过观测天象，即日月星辰的运行规律来确定年、季、月、节气、日、时辰、时刻等时间单位的实践。中国古人通过天文观测制定历法，指导和安排农业生产、祭祀活动、日常生活等。观象授时将时间与空间联系在一起，是古人对天象的观测、认识与应用，对传统人文观念与思想制度产生深刻影响，是中国古代文明的重要基石。

This practice involves observing celestial phenomena, such as the movements of the sun, the moon, and stars, to define temporal units including years, seasons, months, solar terms, days, Chinese zodiac hours and quarter-hours. Ancient Chinese utilized astronomical observations to devise calendars that directed agricultural production, ceremonial rituals, and daily life. This practice connected temporal and spatial dimensions, combining observation and understanding of celestial phenomena with the application of such knowledge. It profoundly influenced traditional cultural concepts and intellectual systems, serving as a fundamental pillar of ancient Chinese civilization.

5. 象天法地
xiàng tiān fǎ dì

Emulating the Patterns of Heaven and Earth

通过对天地形象或者自然现象的模拟，使城市或建筑呈现出形似于天地的布局设计，建立起天地人之间的对应关系。以恒居于天之中的紫微垣作为"天帝"之所，对应都城中"天子"居所，是"象天法地"常取的意象。象天法地常作为中国古代进行城市营建、建筑创作的设计手法，集中反映了中国古人"天人合一"的宇宙观。

This principle involves designing cities or buildings in a way that mirrors the patterns of heaven and earth, or natural phenomena, thus creating a symbolic correspondence between the cosmos, the terrestrial realm, and humanity. A common representation of this principle is seen in how Ziweiyuan (the Celestial Purple Forbidden Enclosure), associated with the abode of Heaven, is mirrored on earth by the residence of the Son of Heaven in the imperial capital. This approach to city construction and architectural design in ancient China encapsulated the cosmological concept of achieving harmony between nature and humanity.

biàn fāng zhèng wèi

6. 辨方正位

Determining Directions and Establishing Proper Positions

《周礼》记"惟王建国，辨方正位"，意思是创建国都，首先要完成对空间的测定，即通过天文观测、大地测量、综合调查，评估一定空间范围的地理区位、资源禀赋以及人文特征等，确定都城的建设位置。元大都营建之初，即通过辨方正位，结合自然环境规划都城布局，并确定中轴线的方位。

As stated in *The Rites of Zhou*, "When a king founds a nation, he must first determine directions and establish proper positions". This entails identifying the most auspicious location for the capital through astronomical observations, geodetic surveys, and comprehensive investigations of the geography, resources, and cultural traits of an area. In the initial planning and construction of the Yuan Dynasty's capital, Dadu, this method was employed to plan the city's layout in harmony with the natural environment and to ascertain the precise orientation of the Central Axis.

yǐ zhōng wéi zūn

7. 以中为尊

According Dignity to the Centre

中国古人在观象授时实践中形成的理念。汉字"中"所象之形表示了辨方正位定时的方法，通过此种方法提供农耕时代最为重要的观象授时服务，是君权产生的基础，并衍生出"中庸""中和""中正""允执厥中""居中而治"等理念。以中为尊，彰显了古代时空观，是中国古代都城营建、城市规划和建筑布局的重要原则。

The ancient Chinese came to see the centre as the place of dignity when they observed the sky to determine the time. The Chinese character for "centre" represents the method of determining directions and establishing proper positions. This method made it possible to provide the most important service in the agricultural age by observing the sky to determine and tell the seasons. It formed the basis of the legitimacy of the monarchy. Other concepts of traditional Chinese thought were also derived from this practice, such as the "mean", "balanced harmony", "exactitude", "holding fast the golden mean", and "governing from a position at the centre". According dignity to the centre is thus an ancient element in the Chinese view of space and time and it was an important principle in capital city planning and construction, as well as architectural layout.

jiànjí suíyóu

8. 建极绥猷

Establishing the National Order and Implementing a Governance of Edification

故宫太和殿内所悬匾额，指君王建立国家秩序、推行教化之治的责任。建极，本意为立表测影、辨方正位定时的方法，引申为立中、建立国家秩序。绥猷，指顺应天道、推行教化之治。建极绥猷显示了天子权力的来源和权力施用的方向。

The inscription *jianji suiyou* (建极绥猷) on the plaque in the Hall of Supreme Harmony in the Forbidden City embodies the emperor's imperative to establish national order and foster governance based on culture and education. *Jianji* (建极) originally refers to the method of erecting a standard for shadow measurement to ascertain direction and time, which is later metaphorically extended to signify the establishment of the central order of the nation. *Suiyou* (绥猷) implies adherence to the natural order and the implementation of an enlightening governance. Collectively, they denote the source of imperial power and the direction of its exercise.

yǔnzhí juézhōng

9. 允执厥中

Holding Fast the Golden Mean

故宫中和殿内所悬匾额，指天子恪守"中"所代表的辨方正位定时之法，秉持中正仁和的理念治理国家的责任。允执，即切实地执行；厥中，即此中，本义是辨方正位定时，引申为中正之道。

The inscription *yunzhi juezhong* (允执厥中) on the plaque in the Hall of Central Harmony of the Forbidden City encapsulates the emperor's duty to embody the principle of centrality, that is, holding fast to the golden mean. This principle is integral to ascertaining proper direction and timing, which are key to governing the nation with a philosophy that emphasizes centrality, righteousness, benevolence, and harmony. *Yunzhi* (允执) signifies the diligent execution of these duties, while *juezhong* (厥中) refers to the essential quality of centrality, denoting the physical act of locating the correct position and timing, and further symbolizing the moral imperative to maintain centrality and righteousness in statecraft.

huáng jiàn yǒu jí
10. 皇建有极

Extensively Establishing the National Order and Legal System

故宫保和殿内所悬匾额,指君王治国理政时,须建极、立中,恪守中正之道的责任。皇,盛大状;皇建即广泛、全面地建立。有极,即有中,引申为国家秩序与法制。"建极绥猷""允执厥中""皇建有极"均体现了"中"在国家治理中的重要意义。

The inscription *huang jian you ji* (皇建有极) on the plaque within the Hall of Preserving Harmony in the Forbidden City refers to the emperor's responsibility to establish the ultimate order and to steadfastly adhere to the path of righteousness when governing the nation. *Huang* (皇) *jian* (建) implies establishing extensively and comprehensively. *You* (有) *ji* (极) means having a central order, which, by extension, refers to the national order and legal system. *Jianji suiyou, yunzhi juezhong,* and *huang jian you ji* all emphasize the significance of the principle of centrality in state governance.

lǐ
11. 礼

Li

社会秩序的总称，用以规范个人与他人、与天地万物之间的关系。"礼"通过对各种有关器物、仪式、制度、建筑等的规定，明确了个人的特定身份及相应的权力和责任，从而区别个人在社会群体中的长幼、亲疏和尊卑等。礼制是关于"礼"的规定，有助于建立社会秩序和维护社会安定。传统中轴线见证了中国古代城市规划建设中蕴含的礼制思想。

Li constitutes a concept that governs public order, structured to orchestrate interactions between individuals, their peers, and the natural world. Through detailed stipulations on various objects, ceremonies, institutions, and structures, *li* delineates specific identities and the corresponding rights and responsibilities, distinguishing social status based on age, kinship, and rank within the community. The ritual system, which encompasses the rules of *li*, aids in establishing public order and ensuring stability. The Traditional Central Axis bears witness to the ritualistic ideas embedded within ancient Chinese urban planning and construction.

yuè
12. 乐

Yue

古代六艺之一，常与"礼"并称，广义上的"乐"指音乐、舞蹈、诗歌等。"乐"有助于人达到平和中正的状态，使人的言行自觉符合礼的要求，从而实现人与人之间的和谐共处。

One of the Six Arts in ancient China, often mentioned in conjunction with *li*, *yue* broadly encompasses music, dance, and poetry. *Yue* facilitates a state of equanimity and balance, encouraging individuals to act in accordance with the principles of *li*, thus fostering harmonious relationships among individuals.

13. 礼乐相成

Li and *Yue* Working in Synergy

  礼和乐两者相辅相成，构成了一种完整的礼仪制度，将阴与阳、秩序与变化、理性与感性有机统一起来，共同维系人伦秩序、维护社会稳定。

Li and *yue* are interdependent, comprising a holistic system of rites that organically integrates opposites such as yin and yang, order and change, and reason and emotion, collectively preserving the natural order of human relationships and social stability.

zhōnghé zhī měi

14. 中和之美

The Beauty of Harmony and Equilibrium

一种整体协调、不偏不倚、和谐平衡的审美感受。在中国传统美学中，中和之美是中国传统文化精神的重要特质，注重整体和谐，崇尚和而不同，强调尊重自然、天人合一。传统中轴线格局对称、左右均衡，在严整的秩序中蕴含着丰富变化，彼此互有制约又相辅相成，体现了中和之美。

This aesthetic perception values overall coordination, impartiality and harmonious balance. In the realm of traditional Chinese aesthetics, the beauty of harmony and equilibrium represents a quintessential aspect of the Chinese cultural ethos, which prizes holistic harmony, celebrates diversity within harmony, and emphasizes respect for nature, along with the harmony between humanity and the natural world. The Traditional Central Axis, with its symmetric pattern and balanced flanks, embodies an orderly structure that accommodates rich variation, where elements mutually constrain and complement one another, manifesting the beauty of harmony and equilibrium.

15. 祭礼
jìlǐ

Sacrificial Ceremony

祭祀之礼，是人伦生活中的一项重要礼仪。依据古礼的规定，祭祀的对象包括天地、日月、山河、海洋以及先人等。人们通过祭礼表达对受祭对象的敬重或畏惧，并希望受祭对象认可自身的言行，从而获得其护佑并享有其赐予的福禄。祭礼对祭祀者及相应仪节的规定，也彰显着祭祀者在人伦生活中的身份与地位。天坛、地坛、日坛、月坛、先农坛、先蚕坛、天神坛、地祇坛和社稷坛等承担着不同的祭礼活动。

The sacrificial ceremony was one of the major rituals in the life of the Chinese. In ancient times, sacrificial ceremonies were held to show reverence for heaven, the earth, the sun, the moon, mountains, rivers, and ancestors. By holding a sacrificial ceremony in this regard, people expressed their respect and awe and prayed for protection and blessings. The rites observed during the sacrificial ceremony reflected the social status of individual participants. The Temple of Heaven, the Altar of Earth, the Altar of the Sun, the Altar of the Moon, the Altar of Agriculture, the Altar of the Goddess of Silkworms, the Altar of Celestial Gods, the Altar of Terrestrial Deities, and the Altar of Land and Grain served different ceremonial activities.

《Kǎogōng Jì》

16.《考工记》

Kaogongji (Book of Diverse Crafts)

汉代补入儒家经典《周礼》并成为其六个篇章之一的著作。《考工记》中关于"匠人建国""匠人营国"的论述涉及都城选址、形制规模、布局结构等方面，描绘了中国理想都城秩序的范式，对中国古代甚至现代城市规划建设都产生了重要影响。

Kaogongji (Book of Diverse Crafts) is one of the six chapters of the Confucian classic *The Rites of Zhou*. The text discusses the roles of craftsmen in the selection of a site for the capital city, its form, scale, layout, and structure. The book outlines an ideal model for the orderly planning of capital cities, profoundly influencing urban planning and construction in ancient China. Its principles continue to echo in modern practices.

zuǒ zǔ yòu shè
17. 左祖右社

Ancestral Temple on the Left and Altar of Land and Grain on the Right

出自《考工记》。指以朝南为正向，在宫城左（东）侧设太庙以祭祀祖先，在宫城右（西）侧设社稷坛以祭祀五土神、五谷神。左祖右社，是中国传统理想都城的营建法则之一。

Originally taken from *Kaogongji*, this concept prescribes that from the perspective of a south-facing palace, a temple for ancestor worship shall be built on the left (east), and an altar for the worship of the deities of land and grain shall be built on the right (west) side. This is one of the principles for planning traditional ideal capital cities in China.

miàn cháo hòushì

18. 面朝后市

Palace in the Front and Market in the Back

出自《考工记》。指以朝南为正向，皇帝处理朝政的宫殿在前（南），百姓买卖商品的市场在后（北）。面朝后市，是中国传统理想都城的营建法则之一。

Originally taken from *Kaogongji*, this is the notion that from the perspective of a south-facing emperor, his palace where he conducted affairs of state was in front, while the marketplace where the people bought and sold goods was behind, north of the palace. It is one of the principles for planning traditional ideal capital cities in China.

chuántǒng guīzhì

19. 传统规制

Traditional Specifications

传统建筑在形制规模、用材用料等级方面的规定。中国古代对建筑的屋顶式样、用材等级、面阔间数、台基层数、装饰色彩等诸多方面都有具体规定。规制代表着建筑的等级和规格，是中国传统礼制的物质表现。如故宫太和殿的重檐庑殿顶、黄琉璃瓦覆顶、面阔十一间、三层汉白玉石雕台基等特征，都是皇家建筑最高规制的体现。

These are guidelines pertaining to the architectural layout, dimensions, and the standard of materials used in traditional building designs. China had specific regulations on many aspects of buildings such as the style of the roof, the grade (quality) of materials, the number of bays, the height of the terrace foundation, and the colours of decorations. Specifications regarding the design and standard of buildings are an expression of China's traditional ritual system. For example, the features of the Forbidden City's Hall of Supreme Harmony, such as a hipped roof with double eaves, a roof topped with yellow glazed tiles, a façade with a width of 11 bays, and a three-level foundation of white marble, are the embodiment of the highest grade of imperial architecture.

móshù
20. 模数

Modularity

在城市规划和建筑中采用的一种标准尺度关系的统称，通过规定典型建筑构件或建筑单元与整体系统的尺度关系，来控制建筑物与建筑群的规模与比例。中国古代的模数制可起到确定城市平面尺度、建筑单体间等级秩序、单体建筑自身内在秩序等重要而独特的作用，且其施用之数呈现出丰富而博大的中国传统数术内涵，尤其表现出与《周易》思想以及中国古代天文、律历的关联。

Modularity in urban planning and architecture refers to the use of a standardized scale relationship to guide design. It involves establishing a scale relationship between typical architectural elements or building units and the overall system to control the size and proportion of both individual buildings and architectural ensembles. The modularity system of ancient China played a crucial and distinctive role in establishing the scale of urban layouts, the hierarchical order among architectural entities, and the internal order within individual buildings. The numerical values applied in this system reflect the rich and expansive essence of traditional Chinese numerology, especially demonstrating its links to the philosophy of *The Book of Changes*, as well as to ancient Chinese astronomical and calendrical science.

第二篇
历史文化要素
Part II
Historic Elements

21. 北京中轴线文化遗产

Běijīng zhōngzhóuxiàn wénhuà yíchǎn

Beijing Central Axis Cultural Heritage

根据《北京中轴线文化遗产保护条例》，北京中轴线文化遗产指北端为北京鼓楼、钟楼，南端为永定门，纵贯北京老城，全长7.8千米，由古代皇家建筑、城市管理设施和居中历史道路、现代公共建筑和公共空间共同构成的城市历史建筑群。2023年1月，中国政府正式向联合国教科文组织申请将"北京中轴线——中国理想都城秩序的杰作"列入《世界遗产名录》。

As stipulated by the *Regulations on the Conservation of Beijing Central Axis Cultural Heritage*, the property encompasses a span from the Drum Tower and Bell Tower at its northern tip to Yongdingmen at the southern limit. It traverses the historical core of Beijing, stretching 7.8 kilometres, and is characterized by an ensemble of imperial buildings, urban management facilities, road remains along the central axis, contemporary public edifices, and public space. In January 2023, the Chinese government formally submitted to UNESCO the dossier of "Beijing Central Axis: A Building Ensemble Exhibiting the Ideal Order of the Chinese Capital" for inscription on the World Heritage List.

Míng-Qīng Běijīngchéng

22. 明清北京城

The City of Beijing in the Ming and Qing Dynasties

明清两代的都城。始建于明洪武元年（公元1368年），在元大都故城基础上改建为北平城，明永乐元年（公元1403年）始称北京。明嘉靖三十二年（公元1553年）添建外城，形成由宫城、皇城、内城、外城组成的四重城垣，整体呈"凸"字形。明清北京城城址位于今北京市区中心，涵盖明清时期北京城护城河及其遗址以内的区域。总面积约63.8平方千米。

Beijing served as the imperial capital during the Ming and Qing dynasties. Starting in 1368, during the reign of Ming Emperor Hongwu, it was restructured from the Yuan Dynasty's capital Dadu, and renamed Beiping. It acquired the name Beijing in 1403, the inaugural year of Ming Emperor Yongle's rule. Later in 1553, under the auspices of Ming Emperor Jiajing, Beijing was further enlarged to encompass an Outer City protruding southwards. The city thus consisted of four walled areas, namely the rectangular Outer City, the square-shaped Inner City, the Imperial City, and the Palace City. With a total area of about 63.8 square kilometres, it locates in present-day central Beijing, covering the area within the old city moat as it existed in the Ming-Qing period.

nèichéng

23. 内城

The Inner City

明清北京城"凸"字形的北部区域。明洪武年间（公元1368年—1398年）、明永乐年间（公元1403年—1424年）在元大都大城基础上改建，设有9个城门，历数十年至明正统年间（公元1436年—1449年）基本成型。明清时期内城中沿皇城四周分布众多衙署、仓廒、寺庙及开放式街坊等，城中商业繁华、民俗活动丰富。内城北至北二环路，南至前三门大街，东西至二环路。总面积约36平方千米。

The Inner City was the square-shaped northern part of the City of Beijing during the Ming and Qing dynasties. With nine gates, it was rebuilt on the site of the Yuan Dynasty's capital Dadu during the reigns of Ming Emperors Hongwu (1368–1398) and Yongle (1403–1424), and basically took its shape during the reign of Ming Emperor Zhengtong (1436–1449) following decades of construction. In the Inner City during the Ming and Qing dynasties, a multitude of government offices, granaries, temples, and open community blocks were situated around the Imperial City, leading to a thriving commercial hub teeming with vibrant folk activities. Covering a total area of about 36 square kilometres, the boundaries of the Inner City reached the present day 2nd Ring Road to the north, to Chongwenmen-Qianmen-Xuanwumen Street to the south, and the 2nd Ring Road on the east and west.

wàichéng

24. 外城

The Outer City

明清北京城"凸"字形的南部区域。明嘉靖三十二年（公元 1553 年）为加强安全防御，在北京内城南面增筑外城，将天坛、先农坛和前三门关厢纳入，设有 7 个城门。外城北至前三门大街，东、西、南至二环路。总面积约 26.7 平方千米。

The Outer City was the rectangular southern part of the City of Beijing in the Ming and Qing dynasties. With seven gates, this part was added to the south of the Inner City, in 1553 during the Ming Emperor Jiajing's reign, to strengthen the security and defence of the city. It incorporated the Temple of Heaven, the Altar of the God of Agriculture, and the areas surrounding Chongwenmen, Qianmen, and Xuanwumen. With a total area of about 26.7 square kilometres, the Outer City covered the area from the present-day Chongwenmen-Qianmen-Xuanwumen Street on the north to the 2nd Ring Road on the east, west, and south.

gōngchéng

25. 宫城

The Palace City

明清两代皇家宫廷所在地，即今北京故宫。明永乐初年在元大都宫城基址上兴建，位置略向南移，建成于永乐十八年（公元 1420 年），位于皇城的核心部位。宫城涵盖明清时期筒子河及其以内的区域，总面积约 1 平方千米，其中宫城城墙内约 0.72 平方千米。

The Palace City, the present-day Forbidden City in Beijing, was the location of the Ming and Qing dynasty palaces. Situated at the heart of the Imperial City, the core structure was established in 1420 on the foundation of the Palace City of Dadu, with minor extension to the south. The Palace City consisted of a walled area with buildings inside, surrounded by a moat (Tongzi River). The total area covered about 1 square kilometre, of which about 0.72 square kilometres were within the walls.

huángchéng

26. 皇城

The Imperial City

内城中围护宫城的皇家领域，曾有皇城墙环绕。始建于元代，明清两代持续发展演变，以宫城为核心，以传统中轴线为骨架，有序布局皇家宫殿、坛庙建筑群、苑囿、衙署库坊、居住宅院。总面积约 6.8 平方千米。

As part of the Inner City that surrounded the Palace City, the Imperial City was once enclosed by its own walls. It was first built in the Yuan Dynasty and continued to be reconstructed and to evolve in the Ming and Qing dynasties. With the Palace City at the core and with the Traditional Central Axis running through this framework, the Imperial City was an orderly layout of royal palaces, temple complexes, gardens, government offices, warehouses and workshops, and residential buildings. The total area of the Imperial City covered approximately 6.8 square kilometres.

sì chóng chéngkuò
27. 四重城廓
The Quadruple-Walled City Layout

  由明清北京城的宫城、皇城、内城、外城四重城垣的轮廓构成的独特城市格局，是老城整体保护的十个重点之一。

The City of Beijing in the Ming and Qing dynasties had a unique configuration in that it comprised the Palace City, the Imperial City, the Inner City, and the Outer City. This quadruple-walled city layout is one of the ten priorities for the comprehensive conservation of the Old City of Beijing.

qípán lùwǎng

28. 棋盘路网

The Chessboard Grid

北京老城呈横平竖直、棋盘式布局的主干道路网，多形成于元代。棋盘路网是北京老城道路网的基本特征，也是老城整体保护的十个重点之一。棋盘路网主要由多条东西向道路和南北向道路组成，东西向道路主要包括平安大街、朝阜路、长安街、前三门大街、两广大街；南北向道路主要包括朝阳门北小街一线、东单北大街一线、王府井大街一线、西单北大街一线、赵登禹路一线。棋盘路网，是中国传统营城理念在北京老城道路建设中的重要体现与传承发展。

The Old City of Beijing features a street system organized in a grid pattern reminiscent of a chessboard, with intersecting main roads running both horizontally and vertically. Established primarily during the Yuan Dynasty, this grid constitutes the essential framework of the city's layout and is highlighted as one of the ten priorities in the comprehensive conservation for Beijing's Old City. The chessboard layout of streets mainly consists of several east-west streets and north-south streets. The main east-west ones are Ping'an Street, Chaofu Road, Chang'anjie Avenue, Chongwenmen-Qianmen-Xuanwumen Street, and Guang'anmen-Guangqumen Street; the main north-south ones are Chaoyangmen North Street, Dongdan North Street, Wangfujing Street, Xidan North Street, and Zhao Dengyu Road. The chessboard road network is an important embodiment, continuation, and development of the traditional Chinese concept of urban planning in Beijing's street layout.

jǐngguān shìláng
29. 景观视廊
Visual Corridors

　　城市中特定观测点与特定建筑或者山林景观等目标物之间形成的视线廊道。传统中轴线上的钟楼、鼓楼、景山万春亭、正阳门箭楼、永定门城楼等景观标志物之间相互眺望形成的视线廊道，凸显出中轴线的空间统领地位，延续着北京老城的营城美学。

Visual corridors refer to the vistas between observation points in a city and landmarks such as specific buildings or hills and trees. The visual corridors between landmark structures along the Traditional Central Axis, like the Bell Tower, the Drum Tower, the Wanchun Pavilion atop Jingshan Hill, the Zhengyangmen arrow tower and the Yongdingmen gate tower, underscore the spatial supremacy of the Central Axis. The visual corridors perpetuate the aesthetic principles of city planning in the Old City of Beijing.

jiēdào duìjǐng

30. 街道对景

Street View Corridors

从城市街道眺望街道尽头特定建筑形成的视线廊道。如前门大街望正阳门箭楼、光明路望天坛祈年殿、永定门内大街望永定门城楼等。

Street View Corridors refer to the view corridors when looking along a street towards specific buildings at the far end. For example, Qianmen Street offers a good view of the Zhengyangmen arrow tower, Guangming Road offers a good view of the Hall of Prayer for Good Harvests at the Temple of Heaven, and Yongdingmen Inner Street offers a good view of the Yongdingmen gate tower.

Zhōng Lóu

31. 钟楼

The Bell Tower

位于传统中轴线最北端，南侧的鼓楼与其共同承担元明清时期的城市计时和报时功能。钟楼始建时由砖石城台、木结构楼阁组成（楼阁于明代烧毁）。清乾隆十二年（公元1747年）在原城台上以砖石结构重建，楼体通高47.9米，是传统中轴线上最高的古代建筑。钟楼通过精巧的设计将声学要求与建筑形式完美融合，加之周围平缓开阔，钟声得以远播。1926年，钟楼内开设电影院。1989年，作为博物馆向公众开放。1996年，公布为全国重点文物保护单位。

Situated at the northern terminus of the Traditional Central Axis, the Bell Tower, together with the Drum Tower to its south, played a pivotal role in timekeeping and time signalling practices during the Yuan, Ming, and Qing dynasties. Originally built with a brick and stone base and topped with a wooden pavilion, the latter was lost to a fire in the Ming Dynasty. In 1747, during Emperor Qianlong's reign, the Qing Dynasty saw the reconstruction of the pavilion atop the existing base, this time using brick and stone for its construction. Standing at 47.9 metres, it is the highest ancient structure along the Traditional Central Axis. The Bell Tower elegantly integrates acoustic functionality with architectural aesthetics, enabling the bell's sound to carry across the expansive, open surroundings. In 1926, the interior of the Bell Tower was used as a cinema. It was opened to the public as a museum in 1989 and was designated a protected site at national level in 1996.

Gǔ Lóu

32. 鼓楼

The Drum Tower

　　位于传统中轴线北端，与其北侧的钟楼共同承担元明清时期的城市计时和报时功能。鼓楼由砖石城台、木结构楼阁组成，楼体内有各类计时装置及 25 面更鼓。每日在预定时间先在鼓楼击鼓定更，之后钟楼撞钟向全城报时，并指挥城门启闭。鼓楼作为传统中轴线北端的眺望点，与景山万春亭的互眺视廊是北京老城最重要的景观视廊之一。1925 年，鼓楼成立京兆通俗教育馆。1987 年，作为博物馆向公众开放。1996 年公布为全国重点文物保护单位。

The Drum Tower, sitting on the Traditional Central Axis just south of the Bell Tower, shared responsibilities for timekeeping and time signalling with the Bell Tower throughout the Yuan, Ming, and Qing dynasties. Comprising a brick and stone base topped with a wooden pavilion, the Drum Tower houses various timekeeping instruments and 25 drums. These drums were played at set times each day to mark the hours, followed by the chimes from the Bell Tower, which resonated across the city, signalling the opening and closing of the city gates. The Drum Tower served as a lookout point at the northern end of the Traditional Central Axis, from where the vista to the Wanchun Pavilion atop Jingshan Hill is one of the most significant scenic views in the Old City of Beijing. In 1925, the Drum Tower was transformed into the Capital Institute for Popular Education. It was opened to the public as a museum in 1987 and was designated a protected site at national level in 1996.

Shíchà Hǎi

33. 什刹海

Shichahai

北京老城内由前海、后海、西海组成的水域。属古高粱河水系，金代称白莲潭，元代称积水潭（是运河漕运的终点），营建元大都时紧傍积水潭东岸选定全城自北而南的中轴线。明代始称什刹海。什刹海作为大运河的组成部分，2013 年公布为全国重点文物保护单位，2014 年列入《世界遗产名录》。

Shichahai is a historic area within the Old City of Beijing, consisting of three interconnected lakes: Qianhai (Front Lake), Houhai (Back Lake), and Xihai (West Lake). These lakes are remnants of the ancient Gaoliang River system. In the Jin Dynasty, the area was known as Bailian Pond. During the Yuan Dynasty, it was called Jishuitan and served as the terminal for the grain transport along the Grand Canal. The central axis of the Dadu was determined along the east bank of Jishuitan. The name Shichahai was first used in the Ming Dynasty. As part of the Grand Canal, Shichahai was designated a protected site at national level in 2013, and in 2014, it was inscribed on the World Heritage List.

Wànníng Qiáo

34. 万宁桥

The Wanning Bridge

位于传统中轴线上、跨玉河而建的石桥。万宁桥建于元至元年间（公元 1264 年—1294 年），为桥闸一体的单孔拱桥，是中轴线上最古老的桥梁，至今仍然在承担着重要的城市交通功能。万宁桥因临近被称为"后门"的地安门，又被称为"后门桥"。元代时，桥西雁翅上的澄清上闸，与下游的澄清中闸和澄清下闸共同配合调节水位，使漕船抵达大运河的终点积水潭（今什刹海）。万宁桥作为大运河的组成部分，2013 年公布为全国重点文物保护单位，2014 年列入《世界遗产名录》。

Spanning the Yuhe River, the stone-made Wanning Bridge is located on the Traditional Central Axis. The bridge was built during the reign of Emperor Zhiyuan (Kublai Khan) between 1264 and 1294 in the Yuan Dynasty, making it the oldest bridge on the central axis. It features a single-arch design with an integrated sluice, which remains vital to the city's traffic today. Due to its proximity to Di'anmen (colloquially known as the Back Gate), it is also referred to as the Back Gate Bridge. During the Yuan Dynasty, the Upper Chengqing Sluice, located on the west side of the bridge and characterized by its wing-like extensions, functioned in tandem with the Middle and Lower Chengqing Sluices further downstream. These structures collectively managed water levels, enabling grain transport boats to navigate successfully to the terminal at Jishuitan, now known as Shichahai. As an integral component of the Grand Canal, the Wanning Bridge was designated a protected site at national level in 2013 and was inscribed on the World Heritage List in 2014.

Dì'ān Mén

35. 地安门

Di'anmen (Gate of Earthly Peace)

明清时期皇城北门，位于传统中轴线上，俗称"后门"。地安门建成于明永乐十八年（公元1420年），原名"北安门"，清顺治八年（公元1651年）改称地安门，以与皇城南门天安门相对。明清两代皇帝北上出行多经地安门。1954年，出于改善交通的目的，地安门拆除。

Di'anmen (Gate of Earthly Peace), the north gate of the Imperial City during the Ming and Qing dynasties on the Traditional Central Axis, was commonly referred to as the Back Gate. Constructed in 1420 during the reign of Ming Emperor Yongle and initially named Bei'anmen (Gate of Northern Peace), it was later renamed Di'anmen in 1651 during the reign of Qing Emperor Shunzhi as a counterpart to the Imperial City's south gate, Tian'anmen (Gate of Heavenly Peace). Emperors of the Ming and Qing dynasties frequently passed through Di'anmen on their journeys northward. In 1954, the gate was dismantled to improve traffic conditions.

Jǐng Shān

36. 景山

Jingshan

明清两代皇家宫苑所在地。景山集高大的山体、华丽的宫殿建筑和秀美的园林建筑于一体。山体堆筑于明永乐年间（公元1403年—1424年），与紫禁城同期营建。明万历（公元1573年—1620年）后期集中营建了山体北面建筑群，使景山成为园居空间。乾隆十六年（公元1751年）在山脊上建成五亭。五亭的建筑形制等级从中央向两侧逐渐降低，凸显出中轴线的空间统领地位。居中的万春亭是传统中轴线的制高点，是登高眺望全城、感受中轴线空间秩序的重要观景点。1928年，景山作为公园向公众开放。2001年公布为全国重点文物保护单位。

Jingshan, the location of the imperial gardens during the Ming and Qing dynasties, comprises a lofty artificial hill, splendid palatial architecture, and elegant garden structures. The formation of the hill occurred during the reign of Ming Emperor Yongle (1403–1424), coinciding with the building of the Forbidden City. In the latter part of the Ming Emperor Wanli's reign (1573–1620), a concerted effort was undertaken to construct a series of buildings to the north of the hill, thus creating a garden-like dwelling space. In 1751, during the reign of Qing Emperor Qianlong, five pavilions were completed atop the ridge. The architectural prominence of these pavilions diminishes from the centre to the sides, reinforcing the preeminent spatial status of the Central Axis. The Wanchun Pavilion, situated at the centre, represents the pinnacle of the Traditional Central Axis. It provides a significant vantage point for overlooking the entire city and experiencing the spatial order of the Central Axis. Jingshan was opened to the public as a park in 1928 and was designated a protected site at national level in 2001.

Shòuhuángdiàn jiànzhùqún

37. 寿皇殿建筑群

The Hall of Imperial Longevity Complex

清代供奉已故帝后御容御像场所。寿皇殿建筑群始建于明万历年间（公元 1573 年—1620 年），原位于景山院墙内东北部，清乾隆十四年（公元 1749 年）拆除并调整基址至中轴线上，仍沿用寿皇殿之名，成为传统中轴线上除故宫以外的第二大建筑群。1956 年，寿皇殿建筑群及其西侧区域由北京市少年宫使用。2013 年，少年宫迁出。2018 年，寿皇殿建筑群作为历史文化展厅向公众开放。

The Hall of Imperial Longevity Complex, used during the Qing Dynasty to enshrine portraits of deceased emperors and empresses, was initially constructed in the northeast section within the walls of Jingshan during the reign of Ming Emperor Wanli (1573–1620). In 1749, during the reign of Qing Emperor Qianlong, it was dismantled, and a new complex bearing the same name was built on the central axis. This made it the second-largest architectural complex on the Traditional Central Axis, following the Forbidden City. Since 1956, the complex and its adjacent western area served as the premise of Beijing Children's Palace, until it moved out in 2013. In 2018, the Complex was opened to the public, offering exhibitions on its history and culture.

Běishàng Mén

38. 北上门

Beishangmen (North Ascending Gate)

原位于传统中轴线上、故宫神武门与景山万岁门（今景山南门）之间的建筑。北上门始建于明代，是从北面进入宫城的必经之路。北上门属于皇城门禁体系，与北上西门、北上东门、景山万岁门及相连墙体所围合的空间共同构成护卫宫城的缓冲地带，以达到增强戒备、整肃皇城的效果。1956年，出于改善交通的目的，北上门拆除。

Beishangmen (North Ascending Gate), situated on the Traditional Central Axis and between the Forbidden City's North Gate (Gate of Divine Prowess) and Jingshan's South Gate (Gate of Longevity), was first constructed during the Ming Dynasty. It served as a crucial passageway for entering the Imperial City from the north. Beishangmen was part of the Imperial City's access control system, which, together with the Beishang West and East Gates, the Gate of Longevity, and the surrounding walls, formed a buffer zone that enhanced security and maintained the solemnity of the Imperial City. In 1956, Beishangmen was dismantled to improve traffic conditions.

Gùgōng
39. 故宫
The Forbidden City

明清两代皇家宫廷所在地。故宫建成于明永乐十八年（公元 1420 年），明清时期称紫禁城，是传统中轴线上规模最大的建筑群，其与北京老城的位置关系反映出中国古代都城"择中立宫"的规划理念。故宫城垣内面积约 72 万平方米，是世界上规模最大、保存最完整的木结构宫殿建筑群，建筑整体依中轴线对称布局，景观秩序严整而富有变化，展现了明清皇家建筑艺术的最高成就。1925 年，故宫博物院成立并向公众开放。1961 年公布为全国重点文物保护单位，1987 年列入《世界遗产名录》。

The Forbidden City served as the imperial residence during the Ming and Qing dynasties. Completed in 1420 during the reign of Ming Emperor Yongle, it was known as the Purple Forbidden City during those times. It is the largest architectural complex on the Traditional Central Axis, and its relationship with the Old City of Beijing exemplifies the ancient Chinese urban planning principle of situating the imperial palace at the city's centre. Covering approximately 720,000 square metres within its walls, the Forbidden City is the world's most extensive and best-preserved wooden structure palace complex. The architecture, symmetrically laid out along the central axis, displays a disciplined and diverse landscape order, showcasing the zenith of imperial architectural artistry from the Ming and Qing dynasties. The Palace Museum was established based on it and opened to the public in 1925. It was designated a protected site at national level in 1961 and inscribed on the World Heritage List in 1987.

Duān Mén

40. 端门

Duanmen (Gate of Correct Deportment)

位于传统中轴线上、天安门与午门之间，是宫城前的夹层门，明清时期存放皇帝仪仗用品和整顿仪仗的场所。端门建成于明永乐十八年（公元1420年），由城台和城楼组成，建筑形制和体量与天安门基本相同，与天安门、午门共同构成进出宫城前的礼仪空间。1961年，端门作为故宫的组成部分公布为全国重点文物保护单位。1999年，端门经修缮后向公众开放。

Positioned on the Traditional Central Axis and between Tian'anmen and the Meridian Gate, Duanmen (Gate of Correct Deportment) serves as the transitional gateway in front of the palace. During the Ming and Qing dynasties, it functioned as a space for storing imperial ceremonial equipment and preparing for imperial processions. Constructed in 1420 during the reign of Ming Emperor Yongle, the structure comprises a raised platform and a gate tower. Its architectural design and scale are essentially similar to those of Tian'anmen. Together with Tian'anmen and the Meridian Gate, it forms the ritualistic space in front of the palace. In 1961, Duanmen was designated a protected site at national level as part of the Forbidden City. In 1999, it was renovated and opened to the public.

Tài Miào

41. 太庙

The Imperial Ancestral Temple

明清两代皇家祖庙所在地。太庙始建于明永乐十八年（公元 1420 年），是明清时期中国最高等级宗庙祭祀建筑，太庙享殿是现存体量最大的明代木结构殿堂之一。太庙位于故宫东南侧，与社稷坛依中轴线东西对称布局，构成"左祖右社"。1925 年，太庙由故宫博物院接管，后作为其分院开放。1950 年，太庙作为北京市劳动人民文化宫向公众开放。1988 年公布为全国重点文物保护单位。

Built in 1420 during the reign of Ming Emperor Yongle, the Imperial Ancestral Temple was the most prestigious temple for ancestral worship in China during the Ming and Qing dynasties. Its Sacrificial Hall is one of the largest existing wooden structures from the Ming Dynasty. Situated to the southeast of the Forbidden City, it is symmetrically laid out with the Altar of Land and Grain along the east and west sides of the central axis, forming a layout known as "ancestral temple on the left and altar of land and grain on the right". In 1925, the Imperial Ancestral Temple came under the management of the Palace Museum and later operated as a branch of the museum. In 1950, it was opened to the public as the Beijing Working People's Cultural Palace. In 1988, it was designated a protected site at national level.

Shèjì Tán

42. 社稷坛

The Altar of Land and Grain

明清两代皇帝祭祀五土神（"社"）和五谷神（"稷"），祈求国事太平、疆土永固、五谷丰登的场所。社稷坛始建于明永乐十八年（公元 1420 年），位于故宫西南侧，与太庙依中轴线东西对称布局，构成"左祖右社"。社稷是国家与政权的象征，社稷坛祭坛上设五色土以象征国土疆域。1914 年，社稷坛辟为中央公园向公众开放，是北京市内最早转变为城市公园的皇家坛庙之一。1928 年，为纪念孙中山先生，中央公园改名为中山公园。1988 年公布为全国重点文物保护单位。

The Altar of Land and Grain was the sacred site where emperors of the Ming and Qing dynasties offered sacrifices to the God of Land and the God of Grain, in hopes of ensuring national peace, territorial integrity, and agricultural prosperity. Built in 1420 during the reign of Ming Emperor Yongle, it is located to the southwest of the Forbidden City. Symmetrically aligned with the Imperial Ancestral Temple along the west and east sides of the central axis, it forms a layout known as "ancestral temple on the left and altar of land and grain on the right". The Altar symbolizes the state and its governance, with soils of five different colours representing all reaches of the nation's territory. In 1914, the Altar was converted into Central Park, marking one of the first imperial temple-transformed public parks in Beijing. In 1928, it was renamed Zhongshan Park in honour of Dr. Sun Yat-sen (Sun Zhongshan). In 1988, it was designated a protected site at national level.

Tiān'ān Mén

43. 天安门

Tian'anmen (Gate of Heavenly Peace)

明清时期颁布诏令、现代举行重大国事活动和国家庆典的场所。天安门位于传统中轴线上，明永乐十五年（公元1417年）始建时称承天门，清顺治八年（公元1651年）改称天安门，此后成为皇城正南门。天安门由城台和城楼组成，1949年中华人民共和国开国大典在此举行，赋予其重要的象征意义和丰富的时代内涵，是全国各族人民向往的地方。1961年公布为全国重点文物保护单位。

Tian'anmen (Gate of Heavenly Peace) served as the location for the issuance of imperial decrees during the Ming and Qing dynasties, and continues to be the venue for important contemporary national ceremonies and celebrations. Situated on the Traditional Central Axis, it was named Chengtianmen (Gate of Heavenly Mandate) when it was first built in 1417 during the reign of Ming Emperor Yongle. In 1651, during the reign of Qing Emperor Shunzhi, it was renamed Tian'anmen, and thereafter became the official South Gate of the Imperial City. Consisting of a raised platform and a gate tower, Tian'anmen was the venue for the founding ceremony of the People's Republic of China in 1949, endowing it with significant symbolic meaning and rich contemporary connotations. It is a revered site for people of all ethnic groups across the nation. In 1961, it was designated a protected site at national level.

Tiān'ānmén huábiǎo

44. 天安门华表

Tian'anmen Huabiao Columns

位于天安门南北两侧的标志性石柱。天安门华表与天安门同期建设，在天安门内外各设一对，均依中轴线对称分布。华表由汉白玉制成，通高9.57米，柱身雕刻云龙，柱体上方横插云板，柱顶蹲立上古神兽石犼一只，整体端庄秀丽、庄严肃穆，是天安门礼制空间的重要组成部分。1961年，天安门华表作为天安门的组成部分公布为全国重点文物保护单位。

Standing majestically on both the southern and northern sides of Tian'anmen are Huabiao, iconic marble columns erected simultaneously with the construction of Tian'anmen. Each pair of columns is symmetrically placed along the central axis, one pair inside and another pair outside Tian'anmen. Crafted from white marble, these columns rise to a height of 9.57 metres and feature intricate carvings of clouds and dragons. A horizontal cloud board is inserted at the top of the column, above which crouches a stone Hou, an ancient mythical creature. The entire structure is dignified, elegant, solemn, and imposing, serving as an important component of the ceremonial space of Tian'anmen. These columns, as an integral part of the Tian'anmen ensemble, was designated a protected site at national level in 1961.

Tiān'ānmén Guānlǐtái

45. 天安门观礼台

Tian'anmen Reviewing Stands

位于天安门两侧用于重大庆典观礼的构筑物。天安门观礼台于1954年建成，建筑形式和色彩与周围空间环境融为一体，面朝天安门广场设置，可同时容纳21,000人观礼。2019年公布为北京市历史建筑。

Flanking Tian'anmen on both sides are the reviewing stands, constructed for observing grand celebrations. Built in 1954, these stands blend seamlessly with the surrounding environment in terms of architectural form and colour. Positioned to face Tian'anmen Square, they can seat 21,000 spectators. In 2019, they were recognized as Historic Buildings of Beijing.

Nèijīnshuǐ Qiáo

46. 内金水桥

The Inner Golden Water Bridges

位于传统中轴线上、跨内金水河而建的汉白玉石桥。内金水桥位于故宫太和门南侧，随故宫建成于明永乐十八年（公元1420年），为五座并列的单孔拱桥，依中轴线对称分布。内金水桥作为故宫的组成部分，1961年公布为全国重点文物保护单位，1987年列入《世界遗产名录》。

Situated on the Traditional Central Axis, the Inner Golden Water Bridges, crafted from white marble, arch over the Inner Golden Water River. Located just south of the Gate of Supreme Harmony within the Forbidden City, they were completed in 1420 concurrently with the construction of the Forbidden City during the reign of Ming Emperor Yongle. The ensemble consists of five single-arch bridges, symmetrically aligned and distributed along the central axis. As an integral part of the Forbidden City, the Inner Golden Water Bridges were designated a protected site at national level in 1961 and were later inscribed on the World Heritage List in 1987.

Wàijīnshuǐ Qiáo

47. 外金水桥

The Outer Golden Water Bridges

位于传统中轴线上、跨外金水河而建的七座并列的汉白玉石桥。外金水桥建成于明永乐十八年（公元 1420 年），在天安门南侧依中轴线对称分布，桥面宽度、柱头形式和装饰细节的等级由中央向两侧逐渐降低，体现了以中为尊的理念。位于中间的五座桥为三孔拱桥，与天安门的五座券门相对，外侧两座分别与太庙南门和社稷坛南门相对。1961 年，外金水桥作为天安门的组成部分公布为全国重点文物保护单位。

Located on the Traditional Central Axis and spanning the Outer Golden Water River, the Outer Golden Water Bridges consist of seven white marble bridges aligned in parallel. Constructed in 1420 during the reign of Ming Emperor Yongle, these bridges are symmetrically arranged along the central axis, south of Tian'anmen. Their design showcases a hierarchical order, with the width of the bridges, styles of column heads, and intricacies of decorative details diminishing progressively from the centre towards the edges. This arrangement embodies the principle of honouring the centre above all. The central five bridges, designed with three arches each, align with the five arched gateways of Tian'anmen, while the outer two bridges correspond with the south gates of the Imperial Ancestral Temple and the Altar of Land and Grain, respectively. As integral part of the Tian'anmen ensemble, the Outer Golden Water Bridges were designated a protected site at national level in 1961.

Qiānbù Láng

48. 千步廊

The Thousand-Step Galleries

明清时期天安门前依中轴线对称布局的廊庑。千步廊与承天门（后称天安门）、长安左门、长安右门、大明门（后称大清门、中华门）共同围合为封闭的T型广场区域。千步廊在景观上强化了皇城主入口处的空间纵深感，其内部空间则多用于储藏其附近衙门的文书等。千步廊东西两侧集中了六部、五府等行政机构，是明清时期中央政府机关集中办公地。1914年，千步廊被拆除。

During the Ming and Qing dynasties, the Thousand-Step Galleries referred to two rows of buildings symmetrically situated on the east and west sides of the Central Axis, to the south of Tian'anmen. Built in the Ming Dynasty, they formed a closed T-shaped plaza, together with Chengtianmen (later renamed Tian'anmen), Chang'an Left Gate, Chang'an Right Gate, and Daming Gate (renamed Daqing Gate in the Qing Dynasty and again as Zhonghua Gate in the Republic eras). The long corridor between the galleries amplified the spatial depth at the main entrance to the Imperial City. The galleries were typically used to store official documents and other items for adjacent government offices. Key government offices, including the six ministries and five bureaus, and other organizations, were located to the east and west sides of the Thousand-Step Galleries, thus forming a centralized administrative hub during the Ming and Qing periods. The Thousand-Step Galleries were dismantled in 1914.

Tiān'ānmén Guǎngchǎng

49. 天安门广场

Tian'anmen Square

党和国家盛大庆典、重大集会和外事迎宾的重要场所。天安门广场位于传统中轴线的核心位置，原为封闭的皇家宫廷广场，形成于明永乐年间（公元 1403 年—1424 年），自 1911 年起经多次改扩建形成今日的格局。位于广场内的毛主席纪念堂和人民英雄纪念碑，以及广场东西两侧的中国国家博物馆和人民大会堂等现代建筑进一步强化了中轴线的空间统领地位。天安门广场南北长 860 米、东西宽 500 米，气势磅礴、宏伟壮观，是世界上最大的城市中心广场，是中华民族团结和祖国繁荣昌盛的重要象征。

Tian'anmen Square serves as a pivotal venue for grand celebrations, significant gatherings, and the hosting of foreign guests organized by the Communist Party of China and the State. Located at the heart of the Traditional Central Axis, Tian'anmen Square was initially an enclosed palace square developed during the reign of Ming Emperor Yongle (1403–1424). It has undergone several expansions and modifications since 1911, culminating in its current layout. The presence of the Chairman Mao Memorial Hall and the Monument to the People's Heroes within the square, along with modern architectural landmarks such as the National Museum of China and the Great Hall of the People which flank the square, further accentuates the commanding role of the Central Axis in spatial arrangement. Measuring 860 metres from north to south and 500 metres from east to west, the square is both majestic and awe-inspiring, standing as the world's largest urban central square and symbolizing the unity and thriving prosperity of the Chinese nation.

Tiān'ānmén Guǎngchǎng guóqígān
50. 天安门广场国旗杆
The Flagpole on Tian'anmen Square

  天安门广场上举行升降国旗仪式的主要构筑物，是中华人民共和国第一面五星红旗升起的地方。它位于传统中轴线上、天安门广场北端，初立于 1949 年开国大典前夕，最初高 22.5 米，1991 年更换新旗杆时调整为 30 米。

The flagpole on Tian'anmen Square is the principal structure used for national flag-raising and lowering ceremonies, and marks the site where the first national flag of the People's Republic of China was hoisted. Situated on the Traditional Central Axis at the northern edge of Tian'anmen Square, it was erected on the eve of the founding ceremony of the People's Republic of China in 1949. The flagpole originally stood at 22.5 metres in height, which was adjusted to 30 metres in 1991 during the replacement of the flagpole.

Rénmín Yīngxióng Jìniànbēi

51. 人民英雄纪念碑

The Monument to the People's Heroes

　　为纪念1840年以来为了反对内外敌人，争取民族独立和人民自由幸福，在人民解放战争和人民革命等历次斗争中牺牲的人民英雄而修建的纪念碑。人民英雄纪念碑建成于1958年，位于天安门广场中央，是中轴线上第一座现代建筑，采用中国传统碑碣形式，通高38米，碑身正面镌刻有毛泽东主席题写的"人民英雄永垂不朽"八个鎏金大字，碑身背面是毛泽东主席起草、周恩来总理题写的碑文。1961年公布为全国重点文物保护单位。

The Monument to the People's Heroes was erected in honour of the heroes who sacrificed their lives in the struggles from 1840 through the People's Liberation War and the People's Revolution. These battles were fought against both internal and external adversaries, aiming to achieve national independence and to secure the freedom and happiness of the people. Completed in 1958 at the centre of Tian'anmen Square, it is the first modern construction on the central axis. The monument stands 38 metres tall and adopts the traditional Chinese stele form. The front of the monument is inscribed with the eight golden characters "The People's Heroes Are Immortal" written by Chairman Mao Zedong, while the back bears an inscription drafted by Chairman Mao Zedong and written by Premier Zhou Enlai. It was designated a protected site at national level in 1961.

Máo Zhǔxí Jìniàntáng

52. 毛主席纪念堂

Chairman Mao Memorial Hall

以毛泽东同志为核心的党的第一代革命领袖集体的纪念堂。毛主席纪念堂建成于1977年，是中轴线上第二座现代建筑，其主体建筑为柱廊型正方体，44根方形花岗岩石柱环抱外廊，雄伟挺拔，庄严肃穆，具有独特的民族风格。1979年公布为北京市文物保护单位。

The Chairman Mao Memorial Hall serves as a memorial to the first generation of revolutionary leadership of the Communist Party of China, with Chairman Mao Zedong at its core. The hall, completed in 1977, stands as the second modern construction on the central axis. The edifice adopts a colonnade design in a square formation, encircled by 44 square granite pillars. This construction, both imposing and solemn, reflects a distinctive national character. It was designated a protected site at municipal level in 1979.

Zhōngguó Guójiā Bówùguǎn
53. 中国国家博物馆
The National Museum of China

代表国家收藏、研究、展示、阐释中华文化代表性物证的博物馆。中国国家博物馆前身为中国历史博物馆和中国革命博物馆，建成于1959年。中国国家博物馆位于天安门广场东侧，与西侧的人民大会堂依中轴线对称布局，并在建筑高度、体量、风格上与其协调呼应，体现出当代对传统规划理念的尊重与延续。中国国家博物馆的建筑风格体现了中国民族主义建筑风格与西方古典主义建筑样式的融合。2011年，登记为不可移动文物。

The National Museum of China stands as a national institution responsible for collecting, researching, exhibiting, and interpreting artefacts that are representative of Chinese culture. It was formerly known as the Museum of Chinese History and the Museum of the Chinese Revolution, which were built in 1959. Located to the east of Tian'anmen Square, the National Museum is symmetrically arranged with the Great Hall of the People across the central axis, with coordinated and corresponding architectural height, volume, and style, reflecting modern-day respect for and continuation of traditional planning concepts. The museum's architecture reflects a hybrid of Chinese national style and Western classical design. In 2011, the museum was designated an immovable cultural heritage site.

Rénmín Dàhuìtáng

54. 人民大会堂

The Great Hall of the People

中国全国人民代表大会的会议场所和全国人民代表大会常务委员会的办公场所，党、国家和各人民团体举行政治活动的重要场所，中国党和国家领导人和人民群众举行政治、外交、文化活动的场所。人民大会堂建成于1959年，位于天安门广场西侧，与东侧的中国国家博物馆依中轴线对称布局，并在建筑高度、体量、风格上与其协调呼应，体现出当代对传统规划理念的尊重与延续。人民大会堂的建筑风格体现了中国民族主义建筑风格与西方古典主义建筑样式的融合。2019年公布为北京市历史建筑。

The Great Hall of the People is the venue for the National People's Congress sessions and houses the offices of its Standing Committee. It is a significant site for the Party, the State, and people's organizations to hold political activities, and serves as a place for Chinese leaders and the people to engage in political, diplomatic, and cultural activities. Completed in 1959, it is situated to the west of Tian'anmen Square, directly opposite the National Museum of China across the central axis, creating a symmetrical layout. Its architectural height, volume, and style are in harmony with the National Museum of China, demonstrating a modern reverence for and continuation of traditional planning concepts. The Great Hall of the People's architecture reflects a hybrid of Chinese national style and Western classical design. It was designated a Historic Building of Beijing in 2019.

Zhèngyáng Mén

55. 正阳门

Zhengyangmen

明清北京内城正南门，俗称"前门"。正阳门始建于明永乐十七年（公元1419年），至明正统四年（公元1439年）形成城楼、箭楼、瓮城一体的格局，与其南侧护城河及正阳桥共同构成完整的防御体系。正阳门位于传统中轴线上，是明清时期皇帝南郊祭祀和南苑围猎的必经之路，是北京城门中规模最大、等级最高的建筑。正阳门地区逐渐成为城市重要对外交通枢纽，1915年，出于改善交通的目的，拆除正阳门瓮城，改造箭楼并在建筑装饰中融入了西洋元素。1988年公布为全国重点文物保护单位。

Zhengyangmen, commonly known as Qianmen (The Front Gate), served as the South Gate of the Inner City during the Ming and Qing dynasties. Constructed in 1419 during Emperor Yongle's reign and further improved until 1439 during Emperor Zhengtong's reign in the Ming Dynasty, it evolved into a fortified complex. This complex featured a gate tower, an arrow tower, and a barbican, complemented by a defensive moat and the Zhengyang Bridge to its south. Standing on the Traditional Central Axis, Zhengyangmen was the grandest and most important of the city gates. It was a crucial passageway for imperial processions heading to the southern suburbs for sacrificial ceremonies and hunting expeditions. Over time, the area around Zhengyangmen evolved into an essential hub for urban transportation. In 1915, the barbican was dismantled to facilitate traffic, and the arrow tower underwent modifications that included Western architectual elements. It was designated a protected site at national level in 1988.

Zhèngyáng Qiáo

56. 正阳桥

The Zhengyang Bridge

位于传统中轴线上、跨正阳门外护城河而建的石桥。正阳桥始建于明代，曾是中轴线上单体最大的古桥，桥身为三拱券洞结构，桥面用栏板分隔为三幅路面，中间为御路。1919年，出于改善交通的目的，正阳桥桥面降低高度并加宽，此后经多次改造，于20世纪60年代被掩埋于地下。2022年，正阳桥遗址经考古发掘重现。2023年公布为北京市文物保护单位。

Straddling the moat outside Zhengyangmen, the Zhengyang Bridge is a white marble stone bridge situated on the Traditional Central Axis of Beijing. Constructed in the Ming Dynasty, it once stood as the largest bridge on the Central Axis, boasting a three-arched structure. The deck was divided into three lanes by balustrades, with the central lane reserved exclusively for imperial use. In 1919, the deck was lowered and widened to improve traffic flow. It underwent several further modifications in the following years and was eventually buried underground in the 1960s. The Zhengyang Bridge was uncovered through archaeological excavations in 2022, and subsequently designated a protected site at municipal level in 2023.

Zhèngyángqiáo Páilou

57. 正阳桥牌楼

The Zhengyang Bridge Pailou

位于传统中轴线上、前门大街北端的六柱五间五楼式牌楼，俗称"前门五牌楼"。正阳桥牌楼始建于明朝正统四年（公元 1439 年），是正阳门外重要的标志性构筑物。1955 年，出于改善交通的目的，正阳桥牌楼拆除。2008 年，前门大街改为步行街，正阳桥牌楼得以原址重建。

The Zhengyang Bridge Pailou, commonly referred to as the Five-Archway Pailou of Qianmen, is a majestic structure featuring six pillars, five bays, and five roofs. Located at the northern end of Qianmen Street along the Traditional Central Axis, it was initially erected in 1439 during the reign of Ming Emperor Zhengtong, serving as an important iconic landmark outside the Zhengyangmen. In 1955, the archway was dismantled as part of traffic improvement efforts. Following the pedestrianization of Qianmen Street in 2008, the Zhengyang Bridge Pailou was faithfully reconstructed at its original site.

jūzhōng dàolù

58. 居中道路

Central Streets

将中轴线上重要建筑群与空间节点连接为一个整体的南北向城市道路系统。具体包括：连接钟鼓楼与景山的内城段居中道路，即地安门外大街、地安门内大街；连接正阳门与永定门的外城段居中道路，即前门大街、天桥南大街。自元代起，居中道路是生活、民俗、商业、游览等活动的重要空间。外城段居中道路在明清两代是皇帝从宫城前往南郊祭祀和南苑围猎的必经之路。

The Central Streets refer to the north-south urban roadway system that connects important architectural complexes and spatial nodes along the central axis into an integral ensemble. It includes the Inner City section from the Bell and Drum Towers to Jingshan, also known as Di'anmen Outer Street and Di'anmen Inner Street; and the Outer City section from Zhengyangmen to Yongdingmen, commonly referred to as Qianmen Street and Tiantan South Street. Since the Yuan Dynasty, the Central Streets have been a vital space for daily life, folklore, commerce, and tourism. In the Ming and Qing dynasties, the Outer City section of the Central Streets served as the imperial route for emperors travelling to the southern suburbs for sacrificial ceremonies and hunting expeditions in the royal garden of Nanyuan.

Qiánmén Dàjiē

59. 前门大街

Qianmen Street

正阳门至天桥的居中道路。形成于明初，明清及民国时期称正阳门大街，1965年定名为前门大街，长约1.4千米，是明清时期皇帝从宫城前往南郊祭祀和南苑围猎的必经之路。作为进出内城的重要通道，自明代中期始，街道两侧逐渐聚集大量商铺、会馆、集市等，至今仍是北京重要的传统商业街。

Qianmen Street, originating in the early Ming Dynasty, is a pivotal thoroughfare that extends from Zhengyangmen to Tianqiao (Heavenly Bridge). It was referred to as Zhengyangmen Street during the Ming and Qing dynasties, as well as the Republic of China Period. The street was officially renamed Qianmen Street in 1965. Stretching approximately 1.4 kilometres, it served as the principal route for Ming and Qing emperors travelling from the Forbidden City to the southern suburbs for ritual sacrifices and hunting expeditions in Nanyuan. As an important conduit into and out of the Inner City, the street began to attract plenty of shops, guild halls, and marketplaces from the mid-Ming Dynasty onwards. It continues to be recognized as a significant traditional commercial district in Beijing today.

dāngdāngchē

60. 铛铛车

Diang Diang Tram

民国时期北京有轨电车的俗称，因其铜车铃发出的"铛铛"声而得名。铛铛车于 1924 年至 1966 年期间运行，共 7 条线路，曾通行于正阳门至永定门段、鼓楼至地安门段等居中道路上，是当时北京重要的交通工具之一。2008 年，前门大街（正阳门至珠市口段）改造为步行街，铛铛车作为前门大街的特色景观得以恢复。

The Diang Diang Tram was a popular name for the tramway system in Beijing during the Republic of China Period, so called because of the distinctive "diang diang" rings from its copper bell. The trams operated from 1924 to 1966 with 7 lines in total, and were one of the important means of transportation in Beijing at the time. The trams ran along the Central Streets, including from Zhengyangmen to Yongdingmen and from the Drum Tower to Di'anmen. In 2008, Qianmen Street (from Zhengyangmen to Zhushikou section) was transformed into a pedestrian street, allowing the restoration of the Diang Diang Tram as a distinctive feature of Qianmen Street.

zhōngzhóuxiàn nánduàn dàolù yícún

61. 中轴线南段道路遗存

Road Remains of the Southern Section of the Traditional Central Axis

居中道路地下的古代道路遗存，包括珠市口南中轴道路排水沟渠遗存、珠市口南中轴道路路肩及板沟遗存、永定门内中轴历史道路遗存、永定门北侧石板道遗存。它们共同展现出自明代以来不同时期居中道路真实的历史位置、走向和传统工程构造，实证了居中道路未曾中断的沿用历史。2004年，永定门北侧石板道遗存在永定门公园建设过程中被发现，其余三处于2022年配合道路市政工程进行考古发掘时被发现。2023年公布为北京市文物保护单位。

Beneath the Central Streets, there are four archaeological sites of road remains: the drainage ditch south of Zhushikou, the road shoulder and stone plank-covered drains south of Zhushikou, the road remains inside Yongdingmen, and the stone path remains north of Yongdingmen. They collectively showcase the authentic historical location, direction, and traditional engineering structures of the Central Streets as they have been since the Ming Dynasty, evidencing over four centuries of uninterrupted use. The stone path remains north of Yongdingmen were discovered during the construction of Yongdingmen Park in 2004, and the other remnants were unearthed during archaeological excavations related to road construction projects in 2022. In 2023, they were designated a protected site at municipal level.

Tiān Qiáo

62. 天桥

Tianqiao (Heavenly Bridge)

位于传统中轴线上、跨龙须沟而建的汉白玉石桥。天桥始建年代不详，原为单孔拱桥，明清两代多次重修，是皇帝赴天坛、先农坛祭祀的必经之路，桥下为东西走向、承担城市排水功能的郊坛后河"龙须沟"。清末至民国时期，出于改善交通的目的，天桥降拱改造，后逐步拆除。2013年，在天桥原址偏南的道路中央绿化带内营造相应景观。

Tianqiao (Heavenly Bridge) was a white marble stone bridge located on the Traditional Central Axis, spanning Longxugou (Dragon's Moustache Ditch). The exact date of Tianqiao's original construction is unknown. It was initially a single-arch bridge and was rebuilt several times during the Ming and Qing dynasties, serving as part of the imperial route to the Temple of Heaven and the Altar of the God of Agriculture. Below the bridge flowed Longxugou, running east-west as part of the city's drainage system, to the north of the suburban altars. In the late Qing Dynasty and the Republic of China Period, the arch of the bridge was lowered to facilitate traffic flow, and it was eventually dismantled. In 2013, a bridge scene that mirrors the original setting was recreated within the green belt in the middle of the road, slightly south of the original location of the bridge.

Zhèngyáng Qiáo Shūqú Jì Fāngbēi
63. 正阳桥疏渠记方碑
The Stele with Record of Dredging the Channels near the Zhengyang Bridge

清乾隆皇帝御笔记录敕命疏浚天桥南部御路周边沟渠始末的四面方幢石碑。该碑形制仿照燕墩乾隆御制碑，与另一方复刻的燕墩乾隆御制碑，于清乾隆五十六年（公元1791年）依中轴线对称刊立于天桥东南侧和西南侧。目前，两碑原件分别位于东城区红庙街78号院（原为弘济寺）、首都博物馆。1984年公布为北京市文物保护单位。

A quadrilateral stone stele, with inscriptions of the Qing Emperor Qianlong, documents the imperial decree and the process of dredging the canals surrounding the imperial road south of Tianqiao. Its design was modelled after the Yandun Qianlong Imperial Stele. This stele and its counterpart, a replica of the Yandun Qianlong Imperial Stele, were symmetrically positioned along the central axis on the southeast and southwest sides of Tianqiao in 1791, during the reign of Emperor Qianlong. The original steles are currently located at 78 Hongmiao Street in Dongcheng District (formerly Hongji Temple), and at the Capital Museum respectively. In 1984, it was designated a protected site at municipal level.

Tiān Tán

64. 天坛

The Temple of Heaven

明清两代皇帝祭天的场所。天坛始建于明永乐十八年（公元 1420 年），原名"天地坛"，明嘉靖九年（公元 1530 年）改名为"天坛"，增建圜丘坛、皇穹宇，嘉靖十七年（公元 1538 年）改造原有大祀殿为三重檐圆殿大享殿（今祈年殿前身）。天坛位于北京老城外城东南部，与先农坛分列中轴线东西两侧，是中国现存规模最大、保存最为完整的明清皇家祭祀建筑群之一。天坛整体形态呈北圆南方，由内坛与外坛两部分组成。内坛设有祈谷坛、圜丘坛两座祭坛，以及祈年殿、皇乾殿、皇穹宇、丹陛桥、斋宫等建筑。1918 年天坛辟为公园向公众开放，1961 年公布为全国重点文物保护单位，1998 年列入《世界遗产名录》。

The Temple of Heaven was the site where emperors of the Ming and Qing dynasties performed the Heaven-worshipping ceremony. It was initially constructed in 1420 during the reign of Ming Emperor Yongle and originally called the Temple of Heaven and Earth. It was renamed the Temple of Heaven in 1530 during the reign of Ming Emperor Jiajing, with the addition of the Circular Mound Altar and the Imperial Vault of Heaven. In 1538, the previous Grand Sacrificial Hall was transformed into the present three-eaved Hall of Prayer for Good Harvests. Located in the southeastern part of the Outer City of Beijing and symmetrically aligned with the Altar of the God of Agriculture across the central axis, the Temple of Heaven is one of the largest and most intact surviving imperial sacrificial building complexes from the Ming and Qing dynasties. The Temple of Heaven features a two-tiered layout with an inner enclosure and an outer enclosure, both built with circular walls on the north and square walls on the south. In the inner enclosure sit the Altar of Prayer for Grain, the Circular Mound Altar, and structures such as the Hall of Prayer for Good Harvests, Hall of Imperial Zenith, Imperial Vault of Heaven, Danbi Bridge, and the Fasting Palace. In 1918, the Temple of Heaven was converted into a public park. It was recognized as a protected site at national level in 1961 and inscribed on the World Heritage List in 1998.

Xiānnóng Tán

65. 先农坛

Xiannongtan (Altar of the God of Agriculture)

明清两代皇帝祭祀先农、山川、神祇、太岁诸神的场所。先农坛始建于明永乐十八年（公元1420年），原名"山川坛"，位于北京老城外城西南部，与天坛分列中轴线东西两侧。先农坛整体形态呈北圆南方，由内坛与外坛两部分组成。内坛内设有神厨、太岁殿、神仓三组重要建筑群，以及宰牲亭、先农神坛、焚帛炉、具服殿、观耕台和耤田等，外坛内设有神祇坛及庆成宫建筑群。1915年，先农坛外坛辟为公园向公众开放。1936年，北平市立公共体育场（今先农坛体育场前身）于外坛东南角建立。1991年，北京古代建筑博物馆于先农坛内坛成立。2001年公布为全国重点文物保护单位。

Xiannongtan (Altar of the God of Agriculture) refers to a complex of structures where emperors of the Ming and Qing dynasties conducted sacrifices to the God of Agriculture, gods of mountains and rivers, and *taisui*. Built in 1420 during the reign of Ming Emperor Yongle and originally named Shanchuantan (Altar of Mountains and Rivers), it is located in the southwestern part of the Outer City of Beijing, with the Temple of Heaven positioned directly opposite across the central axis. Xiannongtan features a two-tiered layout with an inner enclosure and an outer enclosure, both built with circular walls on the north and square walls on the south. The inner enclosure houses three important complexes: the Divine Kitchen, the Hall of Taisui, and the Divine Granary, as well as structures including the Hall of Sacrificial Slaughter, the Altar of the God of Agriculture proper, the Burning Furnace, the Dressing Hall, the Ploughing Viewing Terrace, and Jitian (the Emperor's own farmland). The outer enclosure includes the Altar of Celestial Gods, the Altar of Terrestrial Deities, and the Palace of Celebrating Fulfilment complex. The outer enclosure was opened as a public park in 1915, and the Beiping Municipal Public Stadium (the precursor of the present Xiannongtan Stadium) was built at its southeast corner in 1936. The Museum of Ancient Chinese Architecture was established within the inner enclosure in 1991. Xiannongtan was designated a protected site at national level in 2001.

66. 天神坛和地祇坛

Tiānshén Tán hé Dìqí Tán

The Altar of Celestial Gods and the Altar of Terrestrial Deities

明清两代皇帝祭祀天地众神的场所。始建于嘉靖十年（公元 1531 年），两坛东西并列于先农坛内坛南门外、外坛墙内，均为方形单层祭坛。东侧为天神坛，设四座青白石龛，分别供奉云、雨、风、雷诸神；西侧为地祇坛，设九座青白石龛，分别供奉五岳、五镇、五山、四海、四渎、京畿名山大川及天下名山大川诸神。

The Altar of Celestial Gods and the Altar of Terrestrial Deities served as the venues for Ming and Qing emperors to offer sacrifices to the myriad gods and deities of heaven and earth. Built in 1530 during Emperor Jiajing's reign, both altars are square, single-level structures situated side by side outside the south gate of the inner enclosure, and within the walls of the outer enclosure of Xiannongtan (Altar of the God of Agriculture). The Altar of Celestial Gods, located on the east side, features four greyish-white marble stone niches dedicated to the gods of clouds, rain, wind, and thunder. Meanwhile the Altar of Terrestrial Deities, located on the west side, has nine greyish-white marble niches for different groups of deities: the Five Yues, the Five Zhens, the Five Mountains with imperial tombs of Ming and Qing dynasties, the Four Seas, the Four Rivers, as well as the renowned mountains and rivers of the capital region and those across the country.

67. 一亩三分地

yī mǔ sān fēn dì

One *Mu* and Three *Fens* of Land

先农坛耤田的代称，因面积为一亩三分而得名。作为祭祀先农的一部分，明清两代皇帝每年农历三月率百官在此举行耕耤礼。皇帝通过亲自耕田并观群臣耕种来完成祭礼，以示对农业的重视，祈求国家农事太平顺利。耤田由皇帝、王公大臣示范性耕种，所产粮食主要供皇家祭祀使用。

One *Mu* and Three *Fens* of Land refers to the Emperor's own farmland at Xiannongtan, which is named for its size according to the traditional Chinese measurements of land. (One *mu* and three *fens* in the Ming and Qing dynasties equalled about 800 square metres.) Serving as an integral part of the agricultural rituals, emperors of the Ming and Qing dynasties would lead officials in a ploughing ceremony here during the third lunar month. By personally ploughing the land and overseeing officials do the same, the emperor underscored the importance of agriculture and sought blessings for a peaceful year and a bountiful harvest countrywide. This piece of farmland was cultivated exemplarily by emperors, princes, and court officials, with the harvested grain primarily used for ritual offerings.

Yǒngdìng Mén

68. 永定门

Yongdingmen

    明清北京外城正南门，是传统中轴线的南端点。永定门始建于明嘉靖三十二年（公元 1553 年），由城楼、箭楼、瓮城三部分组成，是明清两代城市防御体系的重要组成部分，也是明清时期皇帝南苑围猎的必经之路。1949 年 2 月 3 日中国人民解放军由永定门进入北平市区，标志着北平和平解放。20 世纪 50 年代，出于改善交通的目的，永定门瓮城、箭楼、城楼陆续拆除。2005 年，永定门城楼原址重建，永定门箭楼和瓮城以地面铺装形式呈现。

Yongdingmen was the principal south gate of Beijing's Outer City during the Ming and Qing dynasties and served as the southern terminus of the Traditional Central Axis. It was first built in 1553 during the reign of Ming Emperor Jiajing and comprised a gate tower, an arrow tower, and a barbican, playing a vital role in the city's defence system. The gate also functioned as the customary route for emperors' hunting expeditions at Nanyuan. The entry of the Chinese People's Liberation Army through Yongdingmen at February 3, 1949, marked the peaceful liberation of Beiping (now Beijing). In the 1950s, the structures were dismantled to facilitate traffic. In 2005, the gate tower was reconstructed at its original location, and the arrow tower and the barbican were delineated with ground paving.

Yàndūn

69. 燕墩

Yandun

位于传统中轴线南端点永定门城楼西南方向的正方形墩台。始建于元代，是明清北京城外南侧的地标性建筑，具有类似烽火台的预警功能。台顶正中石坛上矗立四面方幢石碑一座，为清乾隆十八年（公元1753年）所立，碑身为乾隆御笔的《帝都篇》《皇都篇》，赞美了北京的山水形胜、国泰民安。1984年公布为北京市文物保护单位。

Yandun, a square platform, is located southwest of Yongdingmen. Built during the Yuan Dynasty, it stood as a landmark south of Beijing city during the Ming and Qing dynasties, functioning similarly to a beacon tower for early warning. Atop the platform, a quadrangular stele was erected in 1753 during the reign of Qing Emperor Qianlong. The two eulogies inscribed on the stele, written by Emperor Qianlong himself, praise the capital city's natural beauty and prosperity, as well as the well-being of the people. It was designated a protected site at municipal level in 1984.

lǎozìhao

70. 老字号

Time-Honoured Brands

历史悠久、拥有世代传承的产品、技艺或服务，具有鲜明的中华民族传统文化背景和深厚的文化底蕴，取得社会广泛认同，形成良好信誉的品牌。北京一些老字号起源于前门大街一带，如全聚德、同仁堂等。

Time-Honoured Brands describe products, skills, or services that have been passed down from generation to generation. They should have a distinctive traditional Chinese cultural background and represent a deep-rooted heritage, be widely recognized by the public and have a fine reputation. Several of Beijing's time-honoured brands, such as Quanjude and Tongrentang, originated in the vicinity of Qianmen Street.

zhōng hé sháo yuè

71. 中和韶乐

Zhong He Shao Yue

明清两代主要用于郊庙祭祀和朝会典礼的皇家宫廷礼仪性音乐。中和韶乐以金、石、丝、竹、土、木、匏、革等 8 种自然材料所制成的笙、瑟、编磬和编钟等 16 种传统乐器演奏，融礼乐歌舞为一体，体现了人与自然和谐相处的礼乐思想。天坛神乐署是明清两代的礼乐学府，演奏中和韶乐的乐舞生都出自神乐署。2021 年，天坛神乐署中和韶乐列入国家级非物质文化遗产代表性项目名录。

Zhong he shao yue (中和韶乐) is the royal music primarily performed for sacrifices at temples and altars, as well as court ceremonies during the Ming and Qing dynasties. Characterized by performances on 16 traditional instruments, such as the *sheng, se, bianqing,* and *bianzhong, zhong he shao yue* utilizes eight natural materials: metal, stone, silk, bamboo, clay, wood, gourd, and leather. This union of rituals, music, songs, and dance exemplifies the ancient Chinese philosophy of harmony between humanity and nature through rituals and music. Traditionally, the performers of *zhong he shao yue*, including musicians and dancers, were trained at the Temple of Heaven's Music Division. In 2021, the *zhong he shao yue* of the Temple of Heaven was included in the National Inventory of Representative Elements of Intangible Cultural Heritage.

bāyì wǔ
72. 八佾舞
Bayi Dance

中国古代规格最高、用于祭祀天地等国家大典的舞蹈。由纵横各8人共64人形成舞列，表演时动作庄严，节奏平稳，无快慢之分，以编磬和编钟等传统乐器伴奏，是传统礼制的重要组成部分。

Bayi Dance, esteemed as the most prestigious in ancient China, was performed during significant national ceremonies, including offerings to heaven and earth. The formation includes sixty-four dancers, arranged in eight rows and eight columns. Their movements are dignified, with a consistent rhythm without any variation in pace. Accompanied by traditional instruments such as the *bianqing* and the *bianzhong*, it serves as a crucial element of the traditional ritual system.

第三篇
世界遗产知识
Part III
Knowledge of World Heritage

73. 文化遗产
wénhuà yíchǎn

Cultural Heritage

人类社会历史发展进程中遗留下来、由人类创造或与人类活动有关的一切有价值的物质和非物质遗存。

All valuable tangible and intangible heritage transmitted from the process of the historical development of human society, created by human beings or related to human activities.

74. 世界遗产
shìjiè yíchǎn

World Heritage

根据《保护世界文化和自然遗产公约》和《实施〈世界遗产公约〉操作指南》，符合突出普遍价值标准，满足完整性和（或）真实性条件，具有良好的保护管理状况，由联合国教科文组织世界遗产委员会认定的具有世界意义的文化和（或）自然遗产。截至 2024 年 6 月，北京是全球范围内世界遗产数量最多的城市，有长城、北京故宫、周口店北京人遗址、颐和园、天坛、明十三陵和大运河等 7 处世界遗产。

World Heritage Sites are cultural and/or natural sites of global significance that are considered by the UNESCO World Heritage Committee as having outstanding universal value, satisfying conditions of integrity and/or authenticity, and having a sound conservation and management system in accordance with the *Convention Concerning the Protection of the World Cultural and Natural Heritage* and its *Operational Guidelines*. As of June 2024, Beijing is the city with the largest number of World Heritage Sites worldwide, with seven World Heritage Sites, namely the Great Wall, the Forbidden City, the Peking Man Site at Zhoukoudian, the Summer Palace, the Temple of Heaven, the Ming Tombs and the Grand Canal.

75. 世界文化遗产
shìjiè wénhuà yíchǎn

World Cultural Heritage

根据《保护世界文化和自然遗产公约》和《实施〈世界遗产公约〉操作指南》，符合文化遗产突出普遍价值标准，满足完整性和真实性条件，具有良好的保护管理状况，由世界遗产委员会认定的具有世界意义的文化遗产。截至2024年6月，中国共有39项世界文化遗产。

World cultural heritage sites are cultural sites of global significance that are considered by the UNESCO World Heritage Committee as having outstanding universal value, satisfying conditions of integrity and authenticity, and having a sound conservation and management system in accordance with the *Convention Concerning the Protection of the World Cultural and Natural Heritage* and its *Operational Guidelines*. As of June 2024, China has 39 world cultural heritage sites.

76. 世界自然遗产

shìjiè zìrán yíchǎn

World Natural Heritage

根据《保护世界文化和自然遗产公约》和《实施〈世界遗产公约〉操作指南》，符合自然遗产突出普遍价值标准，满足完整性条件，具有良好的保护管理状况，由联合国教科文组织世界遗产委员会认定的具有世界意义的自然遗产。截至 2024 年 6 月，中国共有 14 项世界自然遗产。

World natural heritage sites are natural sites of global significance that are considered by the UNESCO World Heritage Committee as having outstanding universal value, satisfying conditions of integrity, and having a sound conservation and management system in accordance with the *Convention Concerning the Protection of the World Cultural and Natural Heritage* and its *Operational Guidelines*. As of June 2024, China has 14 world natural heritage sites.

shìjiè wénhuà hé zìrán hùnhé yíchǎn
77. 世界文化和自然混合遗产
World Mixed Cultural and Natural Heritage

根据《保护世界文化和自然遗产公约》和《实施〈世界遗产公约〉操作指南》，同时符合文化遗产和自然遗产突出普遍价值标准，满足完整性和真实性条件，具有良好的保护管理状况，由联合国教科文组织世界遗产委员会认定的具有世界意义的遗产。截至2024年6月，中国共有4项世界文化和自然遗产。

World mixed cultural and natural heritage sites are sites of global significance that are considered by the UNESCO World Heritage Committee as having outstanding universal value, satisfying conditions of integrity and authenticity, and having a sound conservation and management system in accordance with the *Convention Concerning the Protection of the World Cultural and Natural Heritage* and its *Operational Guidelines*. As of June 2024, China has 4 world mixed cultural and natural heritage sites.

《Shìjiè Yíchǎn Gōngyuē》
78.《世界遗产公约》
World Heritage Convention

全称《保护世界文化和自然遗产公约》，由联合国教科文组织第17届大会于1972年通过，内容涵盖文化和自然遗产定义、文化和自然遗产与国家和国际保护、保护世界文化和自然遗产政府间委员会和基金、国际援助、教育计划等，旨在认定、保护、保存和传承具有"突出普遍价值"的世界文化和自然遗产。截至2024年2月，《世界遗产公约》共有195个缔约国，是全球范围内缔约国最多的法律文件之一。中国于1985年批准加入该公约。

The *World Heritage Convention* is formally known as the *Convention Concerning the Protection of the World Cultural and Natural Heritage*. Adopted by the UNESCO General Conference in 1972, it covers the definition of the cultural and natural heritage, national and international protection of such heritage, the intergovernmental committee and the fund, international assistance, educational programmes etc. The Convention aims at the identification, protection, conservation, presentation and transmission to future generations of cultural and natural heritage of Outstanding Universal Value. With 195 States Parties as of February 2024, the Convention is one of the most ratified legal instruments in the world. China ratified the Convention in 1985.

《Shíshī〈Shìjiè Yíchǎn Gōngyuē〉Cāozuò Zhǐnán》
79.《实施〈世界遗产公约〉操作指南》

Operational Guidelines for the Implementation of the World Heritage Convention

世界遗产委员会落实《世界遗产公约》的指南性文件，明确了将遗产项目列入《世界遗产名录》和《濒危世界遗产名录》、保护和保存世界遗产、批准世界遗产基金用于国际援助，以及动员国家和国际力量支持该公约的操作程序。该文件定期修订以反映遗产领域的新概念、新知识和新经验。

As the guiding document for the implementation of the *World Heritage Convention*, the *Operational Guidelines* set forth the operational procedures for the inscription of properties on the World Heritage List and the List of World Heritage in Danger, the protection and conservation of World Heritage properties, the granting of International Assistance under the World Heritage Fund, and the mobilization of national and international support in favour of the *World Heritage Convention*. The *Operational Guidelines* are periodically revised to reflect new concepts, knowledge or experiences of the world heritage field.

80. 世界遗产委员会

World Heritage Committee

全称"保护世界文化和自然遗产政府间委员会",有21个成员,由两年一度的《世界遗产公约》缔约国大会选举产生。委员会负责落实《世界遗产公约》,包括决定某个项目是否列入或移出《世界遗产名录》,决定某项世界遗产是否列入或移出《濒危世界遗产名录》,审议已列入《世界遗产名录》项目的保护状况报告并要求缔约国在其世界遗产未得到妥善管理时采取行动等。委员会每年召开一次常会,即世界遗产委员会会议(国内也称"世界遗产大会")。根据至少三分之二成员国的要求,委员会可召开特别会议。

The World Heritage Committee is short for the Intergovernmental Committee for the Protection of the World Cultural and Natural Heritage. It is comprised of 21 members elected by the General Assembly of States Parties to the *World Heritage Convention* held every 2 years. The World Heritage Committee is responsible for implementation of the *World Heritage Convention*, including making decisions on whether a property should be inscribed or deleted from the World Heritage List and which properties inscribed on the World Heritage List are to be inscribed on, or removed from the List of World Heritage in Danger, examining the state of conservation of properties inscribed on the World Heritage List and requesting a specific State Party to take action when its world heritage property is not properly managed. The Committee members meet once a year for an ordinary session, namely the World Heritage Committee session. At the request of at least two thirds of its members, the Committee can organize extraordinary sessions.

Shìjiè Yíchǎn Zhōngxīn
81. 世界遗产中心
World Heritage Centre

联合国教科文组织负责协调世界遗产事务的工作机构。自1992年成立以来，世界遗产中心作为世界遗产委员会的秘书处协助其开展工作，主要职能包括组织世界遗产委员会有关会议和缔约国大会，落实上述会议的决定和决议，协调《世界遗产公约》的日常管理等。

The World Heritage Centre is the working organ and coordinator within UNESCO for all matters related to World Heritage. Established in 1992, the Centre serves as the secretariat of the World Heritage Committee to assist its conduct of work. The main tasks of the Centre include the organization of the meetings of the General Assembly and the Committee, the implementation of decisions of the above-mentioned meetings, as well as coordinating the day-to-day management of the *World Heritage Convention*.

82. 国际古迹遗址理事会

International Council on Monuments and Sites (ICOMOS)

由《世界遗产公约》明确、服务世界遗产委员会的三个咨询机构之一。理事会成立于1965年，负责评估申报列入《世界遗产名录》的遗产项目，监测文化遗产的保护状况（并与世界自然保护联盟共同监测文化和自然混合遗产的保护状况），评估缔约国申请国际援助的诉求，为能力建设活动提供智力支持。

ICOMOS is one of the three Advisory Bodies to the World Heritage Committee specified by the *World Heritage Convention*. Established in 1965, ICOMOS is responsible for evaluation of properties nominated for inscription on the World Heritage List, monitoring the state of conservation of World Heritage cultural properties (and jointly monitoring the state of conservation of World Heritage mixed properties with the International Union for Conservation of Nature), reviewing requests for International Assistance submitted by States Parties, and providing intellectual input and support for capacity-building activities.

《Shìjiè Yíchǎn Mínglù》

83.《世界遗产名录》

World Heritage List

收录根据《世界遗产公约》和《实施〈世界遗产公约〉操作指南》，符合突出普遍价值标准，满足完整性和（或）真实性条件，具有良好的保护管理状况，由联合国教科文组织世界遗产委员会认定的具有世界意义的文化和（或）自然遗产。截至 2024 年 6 月，全球共有 168 个国家的 1199 项世界遗产纳入《世界遗产名录》，包括中国 57 项。

The World Heritage List is comprised of cultural and/or natural sites of global significance that are considered by the UNESCO World Heritage Committee as having outstanding universal value, satisfying conditions of integrity and/or authenticity, and having a sound conservation and management system in accordance with the *Convention Concerning the Protection of the World Cultural and Natural Heritage* and its *Operational Guidelines*. As of June 2024, there are 1,199 World Heritage Sites from 168 States Parties, among which 57 are located in China.

84. 突出普遍价值

Outstanding Universal Value (OUV)

非常独特、超越国家范畴、对全人类当代和后世具有共同重要意义的文化和（或）自然重大价值。世界遗产委员会界定了10条评估突出普遍价值以将遗产项目列入《世界遗产名录》的标准，遗产项目还须同时满足完整性和（或）真实性，并具有良好的保护管理状况以确保得到充足保护，才能被视为具有突出普遍价值。

OUV means cultural and/or natural significance which is so exceptional as to transcend national boundaries and to be of common importance for present and future generations of all humanity. The World Heritage Committee has defined 10 criteria for the assessment of OUV for the inscription of properties on the World Heritage List. To be deemed of OUV, properties must also satisfy conditions of integrity and/or authenticity and have a sound conservation and management system to ensure adequate protection.

shìjiè wénhuà yíchǎn jiàzhí biāozhǔn
85. 世界文化遗产价值标准
Criteria for the Assessment of OUV of Cultural Heritage Sites

世界遗产委员会界定了10条评估世界遗产突出普遍价值的标准，其中以下6条标准适用于世界文化遗产：

（i）代表人类创造天赋的杰作；

（ii）体现特定时期或特定文化区域内人类在建筑或技术发展、史迹性艺术、城镇规划或景观设计等方面重大的交流互鉴；

（iii）为延续至今或业已消逝的文明或文化传统提供独有的或至少是罕有的见证；

（iv）是展示人类历史某个（些）重要阶段的某种建筑、建筑或技术组合、景观的杰出例证；

（v）是代表某种（些）文化或者反映人类与环境相互作用（特别是当其因不可逆变化的影响而变得脆弱时）的传统人类居住、土地或海洋传统利用的杰出例证；

（vi）与具有突出普遍重大意义的事件或活态传统、观念、信仰、艺术和文学作品存在直接的或物质的关联（委员会认为，此条标准最好与其他标准一起使用）。

The World Heritage Committee has defined 10 criteria for the assessment of OUV of World Heritage Sites, among which 6 apply to cultural heritage sites:
(i) to represent a masterpiece of human creative genius;
(ii) to exhibit an important interchange of human values, over a span of time or within a cultural area of the world, on developments in architecture or technology, monumental arts, town-planning or landscape design;
(iii) to bear a unique or at least exceptional testimony to a cultural tradition or

to a civilization which is living or which has disappeared;

(iv) to be an outstanding example of a type of building, architectural or technological ensemble or landscape which illustrates (a) significant stage(s) in human history;

(v) to be an outstanding example of a traditional human settlement, land-use, or sea-use which is representative of a culture (or cultures), or human interaction with the environment especially when it has become vulnerable under the impact of irreversible change;

(vi) to be directly or tangibly associated with events or living traditions, with ideas, or with beliefs, with artistic and literary works of outstanding universal significance. (The World Heritage Committee considers that this criterion should preferably be used in conjunction with other criteria).

zhànlüè mùbiāo
86. 战略目标
Strategic Objectives

　　世界遗产委员会为推动《世界遗产公约》实施而设定的 5 个战略目标，即提升《世界遗产名录》可信度，确保世界遗产得到有效保护，推动开展有效的能力建设，通过宣传提升公众对世界遗产的认知、参与和支持，以及强化社区在实施《世界遗产公约》中的作用。战略目标的 5 个关键词均以 C 开头，因此也称"5C 战略"。

Strategic Objectives refer to the five strategic objectives developed by the World Heritage Committee to facilitate the implementation of the *World Heritage Convention*. With five key words starting with the letter "C", the Strategic Objectives are also referred to as "the 5Cs". Namely:
(i) Strengthening the **Credibility** of the World Heritage List;
(ii) Ensuring the effective **Conservation** of World Heritage properties;
(iii) Promoting the development of effective **Capacity** building in States Parties;
(iv) Increasing public awareness, involvement, and support for World Heritage through **Communication**;
(v) Enhancing the role of **Communities** in the implementation of the *World Heritage Convention*.

87. 完整性
wánzhěngxìng

Integrity

文化遗产核心价值、价值载体及其环境等要素应得到完整保护。文化遗产在历史演化过程中形成的包括各个时代特征、具有价值的物质遗存都应得到尊重。

Elements of cultural heritage such as core values, value carriers and their setting should receive comprehensive protection. The valuable material remains of cultural heritage, including the characteristic features of each era, formed during historical evolution, should be respected.

zhēnshíxìng

88. 真实性

Authenticity

  文化遗产本身的形态、工艺、设计及其环境和它所反映的历史、文化、社会等相关信息的真实性。对文化遗产的保护就是保护这些信息及其来源的真实性，与其相关的文化传统的延续也是对真实性的保护。

The authenticity of the cultural heritage means its form, craftsmanship, design and setting and the historical, cultural, social and other related information that it reflects are credible and truthful. Safeguarding cultural heritage means protecting the above-mentioned information and its sources, and ensuring the continuity of the cultural traditions associated with it.

yánxùxìng

89. 延续性

Continuity

对保护对象适合当代社会生活功能的延续。对于具有活态特征的保护对象，应延续原有功能，并保护其具有文化价值的传统生产、生活方式。

Continuity means preserving the functions of the conservation objects in ways suited to modern social life. For such living heritage, its original functions shall be continued and traditional ways of production and life with cultural value shall be protected.

zhěngtǐ bǎohù

90. 整体保护

Comprehensive Conservation

以系统思维对保护对象、依存环境及活态遗产和社区结构等实施全要素保护和全过程保护，实现在保护中发展、在发展中保护。在北京老城整体保护中，除保护文物、历史建筑、历史文化街区外，还要加强对历史格局、景观视廊、城市风貌等的保护。

Comprehensive conservation means adopting a systemic approach to the all-key-factors and whole-process conservation of the elements to be protected, the environment they depend on for survival, the living heritage, and the community fabrics, so as to realize development in the process of conservation and conservation in the process of development. In the comprehensive conservation of Beijing's Old City, in addition to the conservation of historic monuments, historic buildings and conservation areas, the conservation of historic patterns, visual corridors and city features must also be strengthened.

91. 遗产区

yíchǎnqū

Boundaries for Effective Protection

包含某项世界遗产突出普遍价值全部价值特征并能保障遗产项目完整性和（或）真实性的区域。

Boundaries for effective protection are drawn to incorporate all the attributes that convey the Outstanding Universal Value and ensure the integrity and/or authenticity of the property.

huǎnchōngqū
92. 缓冲区
Buffer Zones

在遗产周围划定的保护区域，包含遗产的紧邻区域，重要视廊，以及为遗产及其保护起到支撑作用的区域或者属性特征。缓冲区的使用和开发受相应的法律和（或）既定要求限制，以便为遗产项目提供额外的保护。

Buffer zones are areas surrounding the nominated property, including the immediate setting of the nominated property, important views and other areas or attributes that are functionally important as a support to the property and its protection. The buffer zones have complementary legal and/or customary restrictions placed on its use and development in order to give an added layer of protection to the property.

93. 遗产构成要素
Elements

能体现某项遗产突出普遍价值并得到有效保护的该项遗产构成部分。

Elements are constituting components of a specific property which express its Outstanding Universal Value and are effectively protected.

94. 世界遗产的报告与监测
shìjiè yíchǎn de bàogào yǔ jiāncè

Reporting and Monitoring of the World Heritage Sites

遗产列入《世界遗产名录》后开展的保护工作。主要包括缔约国定期准备关于遗产保护状况和具体保护措施的报告，以及对世界遗产开展定期报告。

Reporting and monitoring of the World Heritage Sites are the protection carried out after the inscription of a property on the World Heritage List. They mainly include States Parties' regular preparation of reports about the state of conservation and the various protection measures put in place at their sites, and periodic reporting of the World Heritage Sites.

95. 反应性监测
Reactive Monitoring

世界遗产中心、联合国教科文组织其他部门和咨询机构向世界遗产委员会报告某项受到威胁的世界遗产的保护状况的行为。因此，当发生可能影响遗产突出普遍价值或保护状况的异常情况或施工活动时，缔约国应提交特定报告和影响研究。反应性监测也适用于考察已列入或拟列入《濒危世界遗产名录》项目的保护状况或拟将某项遗产从《世界遗产名录》最终除名的情形。

Reactive monitoring is the reporting by the Secretariat, other sectors of UNESCO and the Advisory Bodies to the World Heritage Committee on the state of conservation of specific World Heritage properties that are under threat. To this end, the States Parties shall submit specific reports and impact studies each time exceptional circumstances occur or work is undertaken which may have an impact on the Outstanding Universal Value of the property or its state of conservation. Reactive monitoring is also foreseen in reference to properties inscribed, or to be inscribed, on the List of World Heritage in Danger and in the procedures for the eventual deletion of properties from the World Heritage List.

附录 Appendix

附录一 中国历史年代简表
Appendix 1 A Brief Chronology of Chinese History

\multicolumn{3}{|c|}{远古时代 Prehistory}	
\multicolumn{3}{|c|}{夏 Xia Dynasty}	c. 2070 - 1600 BC
\multicolumn{3}{|c|}{商 Shang Dynasty}	1600 - 1046 BC
周 Zhou Dynasty	\multicolumn{2}{l|}{西周 Western Zhou Dynasty}	1046 - 771 BC
	\multicolumn{2}{l|}{东周 Eastern Zhou Dynasty}	770 - 256 BC
	\multicolumn{2}{l|}{　春秋时代 Spring and Autumn Period}	770 - 476 BC
	\multicolumn{2}{l|}{　战国时代 Warring States Period}	475 - 221 BC
\multicolumn{3}{|c|}{秦 Qin Dynasty}	221 - 206 BC
汉 Han Dynasty	\multicolumn{2}{l|}{西汉 Western Han Dynasty}	206 BC-AD 25
	\multicolumn{2}{l|}{东汉 Eastern Han Dynasty}	25 - 220
三国 Three Kingdoms	\multicolumn{2}{l|}{魏 Kingdom of Wei}	220 - 265
	\multicolumn{2}{l|}{蜀 Kingdom of Shu}	221 - 263
	\multicolumn{2}{l|}{吴 Kingdom of Wu}	222 - 280
晋 Jin Dynasty	\multicolumn{2}{l|}{西晋 Western Jin Dynasty}	265 - 317
	\multicolumn{2}{l|}{东晋 Eastern Jin Dynasty}	317 - 420
	\multicolumn{2}{l|}{十六国 Sixteen States}	304 - 439
南北朝 Southern and Northern Dynasties	南朝 Southern Dynasties	宋 Song Dynasty	420 - 479
		齐 Qi Dynasty	479 - 502
		梁 Liang Dynasty	502 - 557
		陈 Chen Dynasty	557 - 589
	北朝 Northern Dynasties	北魏 Northern Wei Dynasty	386 - 534
		东魏 Eastern Wei Dynasty	534 - 550
		北齐 Northern Qi Dynasty	550 - 577
		西魏 Western Wei Dynasty	535 - 556
		北周 Northern Zhou Dynasty	557 - 581

隋 Sui Dynasty		581 - 618
唐 Tang Dynasty		618 - 907
五代十国 Five Dynasties and Ten States	后梁 Later Liang Dynasty	907 - 923
	后唐 Later Tang Dynasty	923 - 936
	后晋 Later Jin Dynasty	936 - 947
	后汉 Later Han Dynasty	947 - 950
	后周 Later Zhou Dynasty	951 - 960
	十国 Ten States	902 - 979
宋 Song Dynasty	北宋 Northern Song Dynasty	960 - 1127
	南宋 Southern Song Dynasty	1127 - 1279
辽 Liao Dynasty		907 - 1125
西夏 Western Xia Dynasty		1038 - 1227
金 Jin Dynasty		1115 - 1234
元 Yuan Dynasty		1206 - 1368
明 Ming Dynasty		1368 - 1644
清 Qing Dynasty		1616 - 1911
中华民国 Republic of China		1912 - 1949

中华人民共和国1949年10月1日成立
People's Republic of China, founded on October 1, 1949

附录二　中轴线演进年表
Appendix 2 A Brief Chronology of the Evolution for the Traditional Central Axis

年份	事件
1267年 （元至元四年）	元大都始建，选定传统中轴线方位
1264年—1294年 （元至元年间）	万宁桥及镇水兽、澄清上闸逐步建成
1368年 （明洪武元年）	明清北京城（时称北平）营造肇始，在元大都故城基础上缩北墙
1403年 （明永乐元年）	北平始称北京
1406年 （明朝永乐四年）	故宫始建
1419年 （明永乐十七年）	扩北京城南墙，形成今北京内城范围； 正阳门始建
1420年 （明永乐十八年）	故宫初步建成； 天安门建成，时称承天门； 地安门建成，时称北安门； 太庙、社稷坛、天坛、先农坛始建
1439年 （明正统四年）	正阳门形成城楼、箭楼、瓮城一体的格局； 正阳桥牌楼始建
1531年 （嘉靖十年）	天神坛和地祇坛始建
1553年 （明嘉靖三十二年）	明北京城添建外城，中轴线向南延伸； 永定门始建
1564年 （明嘉靖四十三年）	明北京城外城城门的瓮城建成，中轴线延伸至永定门，形成7.8千米的格局
1600年 （明万历二十八年）	景山寿皇殿始建
1651年 （清顺治八年）	承天门改称天安门，北安门改称地安门
1747年 （清乾隆十二年）	钟楼重建，并保留至今
1749年 （清乾隆十四年）	原景山寿皇殿拆除，调整基址至中轴线上重建，仍沿用寿皇殿之名

年份	事件
1751年 （乾隆十六年）	景山山体上建成五亭
1753年 （清乾隆十八年）	燕墩上刊立乾隆御制碑
1791年 （清乾隆五十六年）	正阳桥疏渠记方碑刊立
1911年	天安门南侧封闭的皇家宫廷广场开始改建
1914年	社稷坛辟为公园向公众开放
1915年	正阳门瓮城拆除、正阳门箭楼改造； 先农坛外坛辟为公园向公众开放
1918年	天坛辟为公园向公众开放
1919年	正阳桥桥面降低高度并加宽
1924年	北京有轨电车铛铛车开始运行
1925年	故宫博物院成立并向公众开放； 太庙由故宫博物院接管，后作为其分院开放； 鼓楼成立京兆通俗教育馆
1926年	钟楼内开设电影院
1949年	中华人民共和国开国大典在天安门举行； 天安门广场国旗杆竖立
1950年	太庙作为北京市劳动人民文化宫向公众开放
1954年	地安门拆除； 天安门观礼台建成
1955年	景山辟为公园向公众开放； 正阳桥牌楼拆除
1956年	寿皇殿建筑群及其西侧区域由北京市少年宫使用； 北上门拆除
1957年	永定门城楼拆除
1958年	人民英雄纪念碑建成
1959年	中国国家博物馆前身（中国历史博物馆和中国革命博物馆）建成； 人民大会堂建成
1961年	故宫、天安门、人民英雄纪念碑、天坛分别公布为全国重点文物保护单位

年份	事件
1977年	毛主席纪念堂建成，两年后公布为北京市文物保护单位
1984年	正阳桥疏渠记方碑公布为北京市文物保护单位
1987年	故宫列入《世界遗产名录》； 鼓楼作为博物馆向公众开放
1988年	太庙、社稷坛、正阳门分别公布为全国重点文物保护单位
1989年	钟楼作为博物馆向公众开放
1996年	北京鼓楼、钟楼公布为全国重点文物保护单位
1998年	天坛列入《世界遗产名录》
2001年	景山、先农坛分别公布为全国重点文物保护单位
2004年	永定门北侧石板道遗存在永定门公园建设过程中被发现
2005年	永定门城楼原址重建
2008年	前门大街（正阳门至珠市口段）改为步行街，正阳桥牌楼原址重建，正阳门至珠市口段铛铛车恢复
2013年	万宁桥作为大运河的组成部分公布为全国重点文物保护单位； 景山内少年宫迁出
2014年	万宁桥作为大运河的组成部分列入《世界遗产名录》
2019年	人民大会堂、天安门观礼台分别公布为北京市历史建筑
2021年	天坛神乐署中和韶乐列入国家级非物质文化遗产代表性项目名录
2022年	正阳桥遗址经考古发掘重现； 珠市口南中轴道路排水沟渠遗存、珠市口南中轴道路路肩及板沟遗存、永定门内中轴历史道路遗存在配合道路市政工程进行考古发掘时被发现
2023年	正阳桥遗址、中轴线南段道路遗存分别公布为北京市文物保护单位

Appendix 2 A Brief Chronology of the Evolution of the Traditional Central Axis

Year	Event(s)
1267 (4th year of the reign of Yuan Emperor Zhiyuan)	The construction of the Yuan Dynasty's capital, Dadu, was initiated, along with the determination of the Traditional Central Axis.
1264–1294 (reign of Yuan Emperor Zhiyuan)	Phased construction of the Wanning Bridge, the Water-Harnessing Beasts, and the Upper Chengqing Sluice took place.
1368 (1st year of the reign of Ming Emperor Hongwu)	The urban development of Ming and Qing Beijing (formerly known as Beiping) began, with the northern boundary scaled back from that of Dadu.
1403 (1st year of the reign of Ming Emperor Yongle)	Beiping was renamed to Beijing.
1406 (4th year of the reign of Ming Emperor Yongle)	The Forbidden City's construction commenced.
1419 (17th year of the reign of Ming Emperor Yongle)	Beijing's southern wall was reconstructed further to the south, shaping the current Inner City; The construction of Zhengyangmen began.
1420 (18th year of the reign of Ming Emperor Yongle)	The initial completion of the Forbidden City ocurred; Tian'anmen was constructed, originally named Chengtianmen; Di'anmen was constructed, originally named Bei'anmen; The construction of the Imperial Ancestral Temple, the Altar of Land and Grain, the Temple of Heaven, and the Altar of the God of Agriculture began.
1439 (4th year of the reign of Ming Emperor Zhengtong)	The final structural layout of Zhengyangmen, consisting of a gate tower, an arrow tower and a barbican, was established; The construction of the Zhengyang Bridge Pailou commenced.
1531 (10th year of the reign of Ming Emperor Jiajing)	The construction of the Altar of Celestial Gods and the Altar of Terrestrial Deities was initiated.
1553 (32nd year of the reign of Ming Emperor Jiajing)	The Outer City was added to Beijing, extending the Central Axis further south; The construction of Yongdingmen began.
1564 (43rd year of the reign of Ming Emperor Jiajing)	The barbicans of the gates of the Outer City were completed, stretching the Central Axis to Yongdingmen and establishing a 7.8 km long Central Axis.
1600 (28th year of the reign of Ming Emperor Wanli)	The construction of the Hall of Imperial Longevity Complex in Jingshan began.

Year	Event(s)
1651 (8th year of the reign of Qing Emperor Shunzhi)	Chengtianmen was renamed to Tian'anmen, and Bei'anmen to Di'anmen.
1747 (12th year of the reign of Qing Emperor Qianlong)	The Bell Tower was rebuilt and remains intact today.
1749 (14th year of the reign of Qing Emperor Qianlong)	The original Hall of Imperial Longevity Complex in Jingshan was dismantled and subsequently reconstructed on the central axis, retaining its historical name.
1751 (16th year of the reign of Qing Emperor Qianlong)	The construction of five pavilions atop the Jingshan hill commenced.
1753 (18th year of the reign of Qing Emperor Qianlong)	The Qianlong Imperial Stele was installed on Yandun.
1791 (56th year of the reign of Qing Emperor Qianlong)	The Stele with Record of Dredging the Channels near the Zhengyang Bridge was erected.
1911	Modifications to the enclosed imperial palace square south of Tian'anmen began.
1914	The Altar of Land and Grain was converted into a public park.
1915	The barbican of Zhengyangmen was dismantled, and its arrow tower was modified; The outer enclosure of the Altar of the God of Agriculture was transformed into a public park.
1918	The Temple of Heaven was opened as a public park.
1919	Adjustments were made to the height and width of the Zhengyang Bridge.
1924	Beijing's tram service, known as the Diang Diang Tram, was inaugurated.
1925	The Palace Museum was founded and parts of the Forbidden City were opened to the public; The Imperial Ancestral Temple was brought under the Palace Museum's management and later opened as its subsidiary; The Drum Tower was repurposed as the Capital Institute for Popular Education.
1926	A cinema was opened inside the Bell Tower.
1949	The founding ceremony of the People's Republic of China took place at Tian'anmen; The national flagpole was erected at Tian'anmen Square.

Year	Event(s)
1950	The Imperial Ancestral Temple was made accessible to the public as the Beijing Working People's Cultural Palace.
1954	Di'anmen was dismantled; Tian'anmen Reviewing Stands was completed.
1955	Jingshan was transformed into a public park, and the Zhengyang Bridge Pailou was taken down.
1956	The Hall of Imperial Longevity Complex and its adjacent western area were allocated to the Beijing Children's Palace; Beishangmen was dismantled.
1957	The Yongdingmen gate tower was taken down.
1958	The Monument to the People's Heroes was erected.
1959	The Museum of Chinese History and the Museum of the Chinese Revolution, forerunners of the National Museum of China, were built; The Great Hall of the People was completed.
1961	The Forbidden City, Tian'anmen, the Monument to the People's Heroes, and the Temple of Heaven were designated protected sites at national level
1977	The Chairman Mao Memorial Hall was completed, and two years later was designated a protected site at municipal level.
1984	The Stele with Record of Dredging the Channels near the Zhengyang Bridge was designated a protected site at municipal level.
1987	The Forbidden City was inscribed on the World Heritage List; The Drum Tower was opened to the public as a museum
1988	The Imperial Ancestral Temple, Altar of Land and Grain, and Zhengyangmen were designated protected sites at national level.
1989	The Bell Tower was inaugurated as a museum open to the public.
1996	The Drum Tower and Bell Tower were designated a protected site at national level.
1998	The Temple of Heaven was inscribed on the World Heritage List.
2001	Jingshan and Xiannongtan (Altar of the God of Agriculture) were designated protected sites at national level.

Year	Event(s)
2004	The stone path remains north of Yongdingmen were uncovered during the park's development.
2005	The Yongdingmen gate tower was reconstructed on the original site.
2008	The Zhengyangmen-Zhushikou section of Qianmen Street was transformed into a pedestrian zone, and the original Zhengyang Bridge Pailou was rebuilt; The tram line from Zhengyangmen to Zhushikou was reinstated.
2013	The Wanning Bridge designated a protected site at national level as an integral component of the Grand Canal; The Beijing Children's Palace, located inside the Hall of Imperial Longevity Complex, moved out.
2014	The Wanning Bridge was inscribed on the World Heritage List as an integral component of the Grand Canal.
2019	The Great Hall of the People and Tian'anmen Reviewing Stands were designated Historic Buildings of Beijing.
2021	The *Zhong He Shao Yue* of the Temple of Heaven's Music Division was included in the National Inventory of Representative Elements of Intangible Cultural Heritage.
2022	Excavations revealed the site of Zhengyang Bridge; Archaeological investigations during road construction and urban engineering uncovered the drainage ditch south of Zhushikou, the road shoulder and stone plank-covered drains south of Zhushikou, and the road remains inside Yongdingmen.
2023	The Zhengyang Bridge site and the Road Remains of the Southern Section of the Traditional Central Axis were designated protected sites at municipal level.

参考文献　Bibliography

法律、法规等（Laws and Regulations）

[1]　《保护世界文化和自然遗产公约》("Convention Concerning the Protection of the World Cultural and Natural Heritage") 1972

[2]　《保护世界文化和自然遗产政府间委员会议事规则》("Rules of Procedure of the Intergovernmental Committee for the Protection of the World Cultural and Natural Heritage") 2015

[3]　《北京城市总体规划（2016年—2035年）》(*Beijing Master Plan (2016-2035)*) 2017

[4]　《北京历史文化名城保护条例》("Regulations on the Conservation of the Historic City of Beijing") 2021

[5]　《北京中轴线保护管理规划（2022年—2035年）》("Conservation and Management Plan for the Central Axis of Beijing (2022–2035)") 2023

[6]　《北京中轴线风貌管控城市设计导则》("Guidelines on Urban Design and Style Control Along the Central Axis of Beijing") 2021

[7]　《北京中轴线文化遗产保护条例》("Regulations on the Conservation of Beijing Central Axis Cultural Heritage") 2022

[8]　《实施〈世界遗产公约〉操作指南》("Operational Guidelines for the Implementation of the *World Heritage Convention*") 2021

[9]　《首都功能核心区街区保护更新导则》("Guidelines for the Conservation and Renewal of Blocks in the Core Area of the Capital City of Beijing") 2020

[10]　《首都功能核心区控制性详细规划（街区层面）（2018年—2035年）》("Regulatory Plan for the Core Area of the Capital (Block Level) (2018-2035)") 2020

[11]　《中华人民共和国文物保护法》("Law of the People's Republic of China on Protection of Cultural Relics") 2017

著作（Works）

[12]　北京历史文化名城保护委员会办公室. 北京历史文化名城保护关键词（汉英对照）[M]. 北京：外语教学与研究出版社，2022.

Office of the Beijing Historic City Conservation Committee, *Conservation of the Historic City of Beijing: Key Terms* (Chinese-English), Foreign Language Teaching and Research Press, Beijing, 2022.

[13] 郭璐，武廷海．辨方正位 体国经野——《周礼》所见中国古代空间规划体系与技术方法[J]．清华大学学报（哲学社会科学版），2017（6）．

Guo Lu and Wu Tinghai, "Determining Directions and Establishing Proper Positions; Segmenting the Capital City into Various Districts for Residences and Farms: The Spatial Planning System and Techniques of Ancient China as Seen in *The Rites of Zhou*", *Journal of Tsinghua University (Philosophy and Social Sciences Edition)*, 2017, no. 6.

[14] 贺业钜．考工记营国制度研究[M]．北京：中国建筑工业出版社，1985．

He Yeju, *Research on Urban Construction and Planning Systems as Stated in 'Kaogongji'*, China Building Industry Press, Beijing, 1985.

[15] 晋宏逵．明代北京皇城诸内门考[J]．故宫学刊，2016（17）．

Jin Hongkui, "Examination of the Inner Gates of the Imperial City of Beijing During the Ming Dynasty", *Journal of Gugong Studies*, 2016, no. 17.

[16] 克里斯蒂娜·卡梅伦，梅希蒂尔德·罗斯．百川归海：世界遗产公约的诞生和早期发展[M]．(2022)．南京：南京大学出版社，2022．

Christina Cameron and Mechtild Rössler, *Many Voices, One Vision: The Early Years of the World Heritage Convention*, Routledge, 2016.

[17] 孔庆普．中国古桥结构考察[M]．北京：东方出版社，2014．

Kong Qingpu, *Investigation into the Structures of Ancient Chinese Bridges*, Orient Press, Beijing, 2014.

[18] 什刹海研究会，什刹海景区管理处．什刹海志[M]．北京：北京出版社，2003．

Shichahai Research Association and Shichahai Scenic Area Administrative Office, *Shichahai Annals*, Beijing Publishing House, Beijing, 2003.

[19] 唐晓峰．乾隆皇帝与北京城中轴线[J]．北京文史，2023（4）．

Tang Xiaofeng, "Emperor Qianlong and the Central Axis of Beijing", *History of Beijing*, 2023, no. 4.

[20] 王南．元上都、大都与中都规划设计之比例、尺度与数术内涵探析（上）（下）[J]，建筑史学刊，2023（3）（4）．

Wang Nan, "Analysis of Proportions, Scales, and Mathematical Insights in the Planning and Design of Yuan Shangdu, Dadu, and Zhongdu", *Journal of Architectural History*, 2023, no. 4, no. 5.

[21] 徐斌，武廷海，王学荣．秦咸阳规划中象天法地思想初探 [J]. 城市规划，2016（12）．
Xu Bin, Wu Tinghai, and Wang Xuerong, "Exploration of the 'Imitating Heaven and Earth' Concept in Qin's Planning Xianyang", *City Planning Review*, 2016, no. 12.

[22] 张宝秀．元明清时期北京中轴线的起源与发展 [J]．北京文史，2023（4）．
Zhang Baoxiu, "Origins and Evolution of Beijing's Central Axis in the Yuan, Ming, and Qing Dynasties", *History of Beijing*, 2023, no. 4.

[23] 《中华思想文化术语》编委会．中华思想文化术语（汉英对照）[M]. 北京：外语教学与研究出版社，2020.
Editorial Board of *Key Concepts in Chinese Thought and Culture*, *Key Concepts in Chinese Thought and Culture*(Chinese-English), Foreign Language Teaching and Research Press, Beijing, 2020.

索引 Index

B

bāyì wǔ	八佾舞	*Bayi* Dance	076
Běijīng chéngshì zhōngzhóuxiàn	北京城市中轴线	The Central Axis of Beijing City	003
Běijīng zhōngzhóuxiàn wénhuà yíchǎn	北京中轴线文化遗产	Beijing Central Axis Cultural Heritage	025
Běishàng Mén	北上门	Beishangmen (North Ascending Gate)	042
biàn fāng zhèng wèi	辨方正位	Determining Directions and Establishing Proper Positions	008

C

| chuántǒng guīzhì | 传统规制 | Traditional Specifications | 021 |
| chuántǒng zhōngzhóuxiàn | 传统中轴线 | The Traditional Central Axis | 004 |

D

dāngdāngchē	铛铛车	Diang Diang Tram	064
Dì'ān Mén	地安门	Di'anmen (Gate of Earthly Peace)	039
Duān Mén	端门	Duanmen (Gate of Correct Deportment)	044

115

F

| fǎnyìngxìng jiāncè | 反应性监测 | Reactive Monitoring | 102 |

G

gōngchéng	宫城	The Palace City	029
Gǔ Lóu	鼓楼	The Drum Tower	036
guānxiàng-shòushí	观象授时	Observing Celestial Phenomena to Determine Time	006
Gùgōng	故宫	The Forbidden City	043
Guójì Gǔjì Yízhǐ Lǐshìhuì	国际古迹遗址理事会	International Council on Monuments and Sites (ICOMOS)	088

H

huǎnchōngqū	缓冲区	Buffer Zones	099
huáng jiàn yǒu jí	皇建有极	Extensively Establishing the National Order and Legal System	012
huángchéng	皇城	The Imperial City	030

J

jiànjí suíyóu	建极绥猷	Establishing the National Order and Implementing a Governance of Edification	010
jiēdào duìjǐng	街道对景	Street View Corridors	034
jìlǐ	祭礼	Sacrificial Ceremony	017

Jǐng Shān	景山	Jingshan	040
jǐngguān shìláng	景观视廊	Visual Corridors	033
jūzhōng dàolù	居中道路	Central Streets	062

K

《Kǎogōng Jì》	《考工记》	*Kaogongji* (*Book of Diverse Crafts*)	018

L

lǎozìhao	老字号	Time-Honoured Brands	074
lǐ	礼	*Li*	013
lǐyuè xiāngchéng	礼乐相成	*Li* and *Yue* Working in Synergy	015

M

Máo Zhǔxí Jìniàntáng	毛主席纪念堂	Chairman Mao Memorial Hall	056
miàn cháo hòushì	面朝后市	Palace in the Front and Market in the Back	020
Míng-Qīng Běijīngchéng	明清北京城	The City of Beijing in the Ming and Qing Dynasties	026
móshù	模数	Modularity	022

N

nèichéng	内城	The Inner City	027
Nèijīnshuǐ Qiáo	内金水桥	The Inner Golden Water Bridges	050

Q

Qiānbù Láng	千步廊	The Thousand-Step Galleries	052
Qiánmén Dàjiē	前门大街	Qianmen Street	063
qípán lùwǎng	棋盘路网	The Chessboard Grid	032

R

| Rénmín Dàhuìtáng | 人民大会堂 | The Great Hall of the People | 058 |
| Rénmín Yīngxióng Jìniànbēi | 人民英雄纪念碑 | The Monument to the People's Heroes | 055 |

S

Shèjì Tán	社稷坛	The Altar of Land and Grain	046
Shíchà Hǎi	什刹海	Shichahai	037
shìjiè wénhuà hé zìrán hùnhé yíchǎn	世界文化和自然混合遗产	World Mixed Cultural and Natural Heritage	083
shìjiè wénhuà yíchǎn	世界文化遗产	World Cultural Heritage	081
shìjiè wénhuà yíchǎn jiàzhí biāozhǔn	世界文化遗产价值标准	Criteria for the Assessment of OUV of Cultural Heritage Sites	091
shìjiè yíchǎn	世界遗产	World Heritage	080
shìjiè yíchǎn de bàogào yǔ jiāncè	世界遗产的报告与监测	Reporting and Monitoring of the World Heritage Sites	101
《Shìjiè Yíchǎn Gōngyuē》	《世界遗产公约》	*World Heritage Convention*	084
《Shìjiè Yíchǎn Mínglù》	《世界遗产名录》	World Heritage List	089

Shìjiè Yíchǎn Wěiyuánhuì	世界遗产委员会	World Heritage Committee	086
Shìjiè Yíchǎn Zhōngxīn	世界遗产中心	World Heritage Centre	087
shìjiè zìrán yíchǎn	世界自然遗产	World Natural Heritage	082
《Shíshī〈Shìjiè Yíchǎn Gōngyuē〉Cāozuò Zhǐnán》	《实施〈世界遗产公约〉操作指南》	*Operational Guidelines for the Implementation of the World Heritage Convention*	085
Shòuhuángdiàn jiànzhùqún	寿皇殿建筑群	The Hall of Imperial Longevity Complex	041
sì chóng chéngkuò	四重城廓	The Quadruple-Walled City Layout	031

T

Tài Miào	太庙	The Imperial Ancestral Temple	045
Tiān Qiáo	天桥	Tianqiao (Heavenly Bridge)	066
Tiān Tán	天坛	The Temple of Heaven	068
Tiān'ān Mén	天安门	Tian'anmen (Gate of Heavenly Peace)	047
Tiān'ānmén Guǎngchǎng	天安门广场	Tian'anmen Square	053
Tiān'ānmén Guǎngchǎng guóqígān	天安门广场国旗杆	The Flagpole on Tian'anmen Square	054
Tiān'ānmén Guānlǐtái	天安门观礼台	Tian'anmen Reviewing Stands	049
Tiān'ānmén huábiǎo	天安门华表	Tian'anmen Huabiao Columns	048

tiānrén-héyī	天人合一	Harmony Between Nature and Humanity	005
Tiānshén Tán hé Dìqí Tán	天神坛和地祇坛	The Altar of Celestial Gods and the Altar of Terrestrial Deities	070
tūchū pǔbiàn jiàzhí	突出普遍价值	Outstanding Universal Value (OUV)	090

W

wàichéng	外城	The Outer City	028
Wàijīnshuǐ Qiáo	外金水桥	The Outer Golden Water Bridges	051
Wànníng Qiáo	万宁桥	The Wanning Bridge	038
wánzhěngxìng	完整性	Integrity	094
wénhuà yíchǎn	文化遗产	Cultural Heritage	079

X

Xiānnóng Tán	先农坛	Xiannongtan (Altar of the God of Agriculture)	069
xiàng tiān fǎ dì	象天法地	Emulating the Patterns of Heaven and Earth	007

Y

Yàndūn	燕墩	*Yandun*	073
yánxùxìng	延续性	Continuity	096
yī mǔ sān fēn dì	一亩三分地	One *Mu* and Three *Fens* of Land	071

yǐ zhōng wéi zūn	以中为尊	According Dignity to the Centre	009
yíchǎn gòuchéng yàosù	遗产构成要素	Elements	100
yíchǎnqū	遗产区	Boundaries for Effective Protection	098
Yǒngdìng Mén	永定门	Yongdingmen	072
yuè	乐	*Yue*	014
yǔnzhí juézhōng	允执厥中	Holding Fast the Golden Mean	011

Z

zhànlüè mùbiāo	战略目标	Strategic Objectives	093
zhěngtǐ bǎohù	整体保护	Comprehensive Conservation	097
Zhèngyáng Mén	正阳门	Zhengyangmen	059
Zhèngyáng Qiáo	正阳桥	The Zhengyang Bridge	060
Zhèngyáng Qiáo Páilou	正阳桥牌楼	The Zhengyang Bridge Pailou	061
Zhèngyáng Qiáo Shūqú Jì Fāngbēi	正阳桥疏渠记方碑	The Stele with Record of Dredging the Channels near the Zhengyang Bridge	067
zhēnshíxìng	真实性	Authenticity	095
zhōng hé sháo yuè	中和韶乐	*Zhong He Shao Yue*	075
Zhōng Lóu	钟楼	The Bell Tower	035
Zhōngguó Guójiā Bówùguǎn	中国国家博物馆	The National Museum of China	057

zhōnghé zhī měi	中和之美	The Beauty of Harmony and Equilibrium	016
zhōngzhóuxiàn nánduàn dàolù yícún	中轴线南段道路遗存	Road Remains of the Southern Section of the Traditional Central Axis	065
zuǒ zǔ yòu shè	左祖右社	Ancestral Temple on the Left and Altar of Land and Grain on the Right	019